Computerization in Developing Countries

Computerization in Developing Countries
Model and Reality

Per Lind

London and New York

First published 1991
by Routledge
11 New Fetter Lane, London EC4P 4EE

Simultaneously published in the USA and Canada
by Routledge
a division of Routledge, Chapman and Hall, Inc.
29 West 35th Street, New York, NY 10001

© 1991 Per Lind

Laser printed from author's discs by
NWL Editorial Services, Langport, Somerset, England

**Printed and bound in Great Britain by
Billings & Sons Limited, Worcester**

British Library Cataloguing in Publication Data
Lind, Per
    Computerization in developing countries: model and reality
    1. Developing countries.   Computer systems
    I. Title
    004.091724

    ISBN 0–415–03818–9

Library of Congress Cataloging in Publication Data
Lind, Per, 1940–
    Computerization in developing countries: model and reality / by Per Lind.
        p.   cm.
    Includes bibliographical references and index
    ISBN 0–415–03818–9
    1. Computers – Developing countries.   2. Electronic data processing –
Developing countries.   I. Title.
    QA76.L49   1991   90–8753
    004′.09172′6 – dc20   CIP

To Lars who left too early

Contents

Contents

Tables

Tables

Figures

Preface

While we tend to assess the usability of computers by referring to processing speed and data-storing capacity, or to response time and data-presentation facilities, comparatively little thought is focused on those models which are the base for computer programs, models of reality. One can live without looking into this. But one cannot fully understand the grounds for low computer performance in developing countries unless models and their applicability are taken into consideration.

This book is an attempt to point out how computer programs, developed in the more advanced industrialized countries and based on models and conceptions of reality that are prevailing in these countries, tend to be inappropriate under different conditions in developing countries. The reason, so it is argued in the book, is that models do not have the same explanation value in different cultures.

Reality is made up of facts. In countries which share the same values and conceptions and where there is a common language, facts are conveyed, through the language, into words. The words are structured into sentences which in turn constitute the model, a substitute for reality. The word-structures of the model must, however, have something in common with the structure of the facts, and only by an a priori knowledge of the structure of the facts, the structure of the words, i.e. the model, can be made. (If there was no mental insight into what the structure is, a meta-language would be required to make the model.)

Let us illustrate this with an example. The production environment of a factory is characterized by two factors: the factory is equipped with 25-year-old machinery and planning of production is very difficult. These are facts of reality. The language transforms these facts into words: '25', 'year', 'machines', 'planning' and 'difficult'. Arranging these words into a structure that corresponds with the facts gives us '25-year-old machines make planning difficult' (implicitly: because they are no longer reliable). This is a model, and based on this model we may, for example, make a computer program that keeps track of ageing machinery. The crucial point here is that the structure of the words must correspond with the structure of the facts in order to make sense, i.e. we must know, even before we set out to make the model, that 25-year-old machines are unreliable.

This awareness is thus something that belongs to the culture, a kind of collective understanding of technology. Even if we translate the words, which is easily done, into another language in another culture with other values, the model will not have the same meaning unless there is a collective understanding of the structure of the facts. If there is no mental understanding, in our example above, due to other values and knowledge, of 25-year-old machines being unreliable, then the model has no or very little meaning. So the basic question is: what meaning does a model have in another cultural context? The inability of languages to perfectly represent facts of reality has been pointed out before by many others. Claude Bernhard (1812–1878), for example, 'one of the most original thinkers in the whole of nineteenth-century medicine' (Campell, 1989), held that our language is actually just an approximation and even in science it is so inexact that if we lose facts out of sight and cling firmly to words we will soon place ourselves extraneous to reality and Ludwig Wittgenstein (1889–1951) referred to 'language games' as kinds of social activities which convey images of reality. As social activities, however, take place within a cultural reference frame, the usability of the language, i.e. the meaning of words, is dependent on this reference frame.

The reader may notice the influence from Ludwig Wittgenstein and that my thinking is based on my interpretation of his thoughts as they appear in his *Tractatus Logico-Philosophicus*. I have, therefore, not been able to resist the temptation of using one or two articles of that book to introduce the different parts that follow.

A note, finally, to the various terms used in the book to describe different technical, economic and social levels of development: North and South, advanced and less advanced, industrialized and industrializing. The terms are used more or less interchangeably in the book and should be seen as linguistic variations rather than as expression of values.

P.L.

Part I

Computers, models and developing countries

Wir machen uns Bilder der Tatsachen.
(We make to ourselves pictures of facts.)

Das Bild ist eine Modelle der Wirklichkeit.
(The picture is a model of reality.)
                    Wittgenstein: *Tractatus Logico-Philosophicus*

Chapter one

Introduction

As computerization spreads in developing countries the question about its contribution to the development process has become an issue of much concern. The advances in computer technology that are so apparent in the most advanced industrialized countries have not left developing countries unaffected, and a strong wish to take part in a comparable evolution is frequently spelled out as follows:

> It is now an accepted fact that there are various areas in a developing country where the pace of development can be speeded up significantly by the use of computers.
>
> (Gupta, 1981)

> We are many in developing countries who found our hope in the computer revolution. We see in it a way to realize our dream to overcome poverty and mediocrity and to reach a society of equity and welfare.
>
> (Benmokhtar, 1984)

A closer look at the present computer usage in many developing countries reveals, however, a considerable uncertainty as to real achievements and benefits. Parallel to the intensive debate about how computerization can be speeded up in the Third World, there is therefore a somewhat less intensive debate about what hampers the utilization of computers in developing countries.

Different phenomena that may have an impact on computer usage and utilization in Southern contexts have been observed. One phenomenon is the discrepancy between today's computer architecture, designed for fast data communication, on the one hand, and poor data communication facilities in many developing countries on the other hand. Another phenomenon is the fact that computer programs for most applications are based on and developed in accordance with models that originate from a Western view of

3

problems and solutions, and which are not always synonymous with Southern practices and information needs.

The following example from a West African country may illustrate this. A European consultancy firm has been commissioned to implement a new computer-based clearance system at the customs authority. At a ceremony, as the new system is officially handed over to the authority, it turns out, however, that a very specific feature is missing. The feature is an extra key on the keyboard with a very particular function, namely to reduce customs duty by a predefined percentage each time the key is pressed. The function is required on those occasions when the officer on duty finds reasons to reduce the customs duty, calculated by the computer according to formal tariffs. The situation occurs, for example, when the officer receives a sufficiently valuable gift from a client!

The necessity of this feature is very obvious, because without it the customs officers are reluctant to use the system. The consultant team must postpone the opening ceremony until the missing feature has been included.

From this example we learn two things. First, that the way reality works must be reflected in the computer solution. Otherwise the solution is of a cosmetic nature only. Second, even very good computer-based solutions, successfully used in one place, are not necessarily equally applicable in another place; in particular if the two places represent different norms of rationality, as in the example above.

It is this second phenomenon that is the focus of interest in this book, and rephrased into the following question: to what extent can the low utilization of computers and the difficulty in achieving even very moderate objectives be ascribed to the lack of fit between the computer model of an application and the actual reality?

The example above also illustrates very clearly the conflict that appears when two different value systems meet and are to be unified in a computer system. One can find other examples where seemingly identical conditions are in reality so different that one and the same computer solution can not possibly fit into both situations.

Too often this type of phenomenon, related to computerization in developing countries, is debated on a global and general level, with little or no room for local characteristics and features. Also when applicability has been the theme, as in studies by Siffin (1976), Kalman (1982) and Rada (1983), the discussions have largely been held on high and general levels, with little distinction made between different computer applications or between characteristics that separate, in this context, different developing countries from each other.

The effect of these general discussions, as pointed out by Muller (1979) and Pascoe (1978), is that a policy related to computers which is sound and effective in one country may be inappropriate under the social and technical conditions of another country. Therefore, as computer models refer to particular activities and functions within an organization, fit and applicability must be assessed, in the first place, from the level where the activities occur and the functions belong. While the model as such may be discussed on a general level, its applicability can not be regarded in isolation from the environment in question.

Part of this book is therefore a micro-study of a manufacturing company and, in particular, its production system. The company is an Egyptian vehicle manufacturer where computers have been implemented to assist managers in production control. This application represents a common problem in industry for which a variety of computer programs have been developed. However, as the studied Egyptian company is exposed to constraints and contingencies that are partly specific to Egypt, an important question is to what extent do the programs reflect these conditions?

Many enthusiastic computer projects in developing countries have been launched where computer solutions, successfully implemented somewhere else, have failed to perform, often because the actual reality was never fully understood (values, authority patterns, rationality, time concepts, etc.). Starting from reality, in an attempt to understand the application area of the actual company, will therefore enable us to identify major parameters influencing the production system of the company and, hence, also the conditions for the computer.

The reader may find, quite rightly, that a relatively large part of the book deals with Egypt and Egyptian industry. The point made here is that many computer systems, for example for planning and control, in general require considerably more detailed analysis before the degree of resemblance between the computer model and the actual reality can be established. The computerization process in many developing countries would probably be slower if more detailed analysis were done before a computer was installed. This would probably jeopardize vendors' marketing plans. It is most unlikely that it would negatively influence the development process. But it would with certainty influence, positively, the cost/efficiency ratio of computerization.

The book is divided into three parts. The first part gives a general view on the computerization process in developing countries and an attempt is made to characterize some basic types of problems of particular relevance for users in Third World countries. The types of problems that are identified here lead to a discussion about how computer programs are based on models and to what extent the applicability of models is limited. One such model, of industrial production, is presented together with a set of computer programs typically designed to cope with the model.

Part two is devoted to the detailed description of a company in a Third World country. This description will serve as a frame of reference and as an illustration to some real problems encountered in a Third World industry.

In the last part the comparison is made between model and reality and conclusions are made about the applicability of the actual models and computer programs.

Chapter two

Computers in developing countries[1]

Introduction

Two things seem to be characteristic as computers spread into developing countries. One is the high degree of confidence shown by many policy makers about computers and computer technology as a blessing for the Third World, as a Pandora's box full of promises and hopes about contributing to the development process. The other is the almost total absence of a critical debate as to whether and to what extent the computer models, practically always designed and developed in the West, are applicable in Southern contexts.

The very enthusiastic attitude towards computer technology that we meet in developing countries and which is not in proportion to real achievements bears some resemblance to the dominating developing paradigm of the 1950s and 1960s. Here the question of whether a particular technology was applicable in a given situation, and thus had the potential of lifting the level of development, was considered less relevant. It was more or less taken for granted that Western technology was generally applicable.

Backwardness was seen as an original development stage which should be followed by a process that released the forces of modernization and different stages of development were just different points on a scale, ranging from the least developed to the most developed countries. The task of development theory was seen to be observing these differences and putting forward suggestions on how to bridge the gaps. A most efficient way of bridging was for poorer countries to imitate already existing technology from the more advanced countries and thus gradually assume the qualities of the latter.

This modernization model was linked to a more general Western development ideology that has now been largely abandoned due to its inadequacy and inability to explain phenomena in the development process. The 1970s and 1980s have therefore seen newer

7

development models emerge.

In spite of this, arguments belonging to the rhetoric of the modernization model still seem to dominate the computer strategies in many developing countries and be the driving force behind the computerization process.[2] One of the basic characteristics of modern computer technology is its total association with the most advanced industrialized countries, being an integrated part of their scientific and technical development. By imitating this technology many developing countries hope to take part in a comparable technical and economic development.

This imitation process includes not only the skills necessary to handle the equipment, but also the conceptual thinking that surrounds and accompanies the technology. The conceptual thinking – it can for instance be the way of defining a problem or identifying a solution – is not neutral, neither semantically nor culturally, nor from a social, technical or economic point of view, but is closely related to a particular society, or group of societies, sharing a common value system.

From this perspective, transfer of computer technology is also a transfer of models from one society to another, an imitation of values and conceptions. A fundamental issue here is to what extent these computer models and conceptions of what a problem is, what a good solution is, etc., that are valid in one society would also fit in another.

As the use of computers in the Third World is implicitly based on the assumption that models, values, conceptions, etc., are in practice transferable between different kinds of societies, the question of fit is worth exploring. But before we go into this question we need to describe further the computerization process in developing countries, what characterizes this process and, in particular, what symptoms we may find that support our inquiry.

The computerization process

The term computerization used in this chapter refers to an on-going process whereby computers and computer technology are introduced and adopted. A number of aspects can be applied to characterize this process. *Growth* is one such aspect, although the most appropriate measure for growth is unclear. *Actors* involved as promoters for the use of computers is another aspect, *applications and users* yet another.

Growth rates in computerization

By the beginning of the 1970s a number of developing countries had gained considerable experience in the use of computers. For example, the first computer appeared in India in 1960, in Kenya in 1961, in 1962 in Indonesia and Egypt, and in 1965 in Malaysia

The growth rate, rather slow at the beginning, increased during the 1970s and towards 1980 there were 540 computers in India, 420 in Malaysia, 400 in Indonesia, 190 in Egypt, and 85 in Kenya. As a comparison there were 172,000 computers installed in Europe in 1975 and 465,000 in 1980. A rough estimate of the total number of computers worldwide in 1980 was 1,250,000, representing a value of 160–170 billion US dollars. Half of those were installed in the USA.[3]

A difficulty encountered when trying to assess the computerization process is the choice of a relevant measure. Muller and Rayfield (1977) point out that:

> One finds simplistic measures such as the number of computers installed or their values, often presented in terms of original purchase price. Sometimes one finds these figures modified by and adjustment designed to take into account the state of the development of the country based on its per capita GNP. One reason for not dwelling on such measures is that they ignore the question of effectiveness of use and the extent to which the computers are used for high priority needs.

Low effectiveness is a frequently reported problem even though definitions of what can be regarded as acceptable effectiveness are very rare. Maruf (1981), to take one example, showed from the public sector in Malaysia that 80 per cent of all installed computers (48 installations) had a utilization of less than 50 per cent. In a Third World comparison this is not exceptional.

Quoting the number of computers installed without taking into account their effective use may thus be misleading. Looking at computerization from an investment point of view, value is an appropriate measure (as long as one is aware of the effectivity aspect). Attempts have thus been made to compare figures of computerization, on an aggregated level (regions) where value is used instead of numbers Rada (1983, p. 205) reported that:

> In terms of the value of data processing equipment the consulting firm Diebold (Europe) estimates that the United States, Japan and Western Europe accounted for 83 per cent of the world total in 1978. The 17 per cent share held by the rest of the world will have risen only marginally to 20 per cent by 1988. Most of this figure is accounted for by Eastern European countries and by some

developing countries. Large Western banking firms possess more computer power that the whole of India. During the period 1978–1988 the 'gap' in the value of equipment between Western countries and the others is expected to grow by a factor of more than two.

However, it should be noticed that the value of data-processing equipment was distributed differently in 1960 compared to, for instance, 1970 (see Table 2.1). The distribution of value does, however, undergo a significant change (see Table 2.2). Any conclusion based on these figures must also take into account the change in price/performance ratio during the period, as shown in Table 2.3.

What kind of conclusion can be drawn from these tables? Unfortunately, figures related to developing countries are not explicitly shown in Table 2.1 but are included in the figures shown for 'Other Countries'. We may, however, try a qualified guess: Kalman (1984, p. 231) shows, not quite surprisingly, that trade with computers (import and export) grows with growing GNP per capita. Figures for the period 1960–80 show that most of the oil exporting countries, the 'newly industrialized countries' and the East European countries have the highest growth rate in GNP whereas the least developed and most other developing countries had a significantly lower GNP growth per capita. It can then be repeated that in the growth in value of data-processing equipment between 1960 and 1980, the least developed and developing countries have a very small share of this computerization. But even the technically more advanced developing countries are far behind the most advanced developed countries. Jamin found that:

> Of importance is the difference in computerization level among the developing countries, levels which vary from beginners to the more advanced countries. According to a survey by the Massachusetts Institute of Technology only 26 developing countries have reached a computerization level comparable to that of West Germany in 1965. Among these countries can be found Chile, Colombia, Malaysia, Peru, The Philippines, Korea, Singapore, Taiwan, Argentina, India, Mexico and Brazil. The majority of the developing countries are still in the very beginning of adopting computer technology.

> (Jamin, 1984, p. 13)

For the five ASEAN countries, Malaysia, Singapore, Thailand, the Philippines and Indonesia, growth in computerization has been substantial (see Table 2.4). Towards the middle of the 1980s, however, this growth rate began to slow down. Stagnation in economic growth was considered a major reason for this. In a recent study, however,

Table 2.1 Distribution of value of data-processing equipment (in Swiss francs – figures in brackets are percentages)

	1960	1970	1973	1978	1983	1988
USA	5.5 (70)	57.9 (62)	77.6 (53)	121 (46)	189 (43)	252 (41)
Western Europe	1.6 (20)	25.3 (27)	40.8 (28)	78 (29)	140 (32)	200 (32)
Japan	0.3 (4)	4.7 (5)	10.5 (7)	21 (8)	31 (7)	48 (8)
Other Countries	0.5 (6)	6.0 (6)	18.4 (12)	45 (17)	80 (18)	122 (19)
Total	7.9 (100)	93.9 (100)	147.3 (100)	265 (100)	440 (100)	622 (100)

Source: Rada/Diebold

Table 2.2 Distribution of value by type of equipment (per cent)

	1960	1970	1973	1978	1983	1988
Computers	75	63	58	43	35	31
Peripherals	22	26	30	40	40	38
Transmission	0.6	7.5	9	15	24	30

Source: Rada/Diebold

Table 2.3 Price/performance change

	Cost in 1975 as % of cost in 1960	Cost in 1985 as % of cost in 1975
Computers	0.5	20
Peripherals (mass storage)	2	10
Communication (line cost)	61	53

Source: Datamation

the authors suggest another explanation for this declining trend:

> We suggest that the current state of computer technology (hardware, software and services), its lack of appropriateness for small business firms and other development-oriented enterprises, and the internal constraints of the informational structure of those firms are placing a ceiling on the potential population size of computer users. In some cases, the existing information structure needs reorganization before computers can be used effectively. In other cases, the available packages of computer systems are un-

Table 2.4 Computerization in five ASEAN countries

	Computers acquired in the year	Cumulative total	Increase in total (%)
1975	63	148	74
1976	89	237	60
1977	120	357	51
1978	220	577	62
1979	343	920	59
1980	556	1,476	60
1981	748	2,224	51
1982	833	3,057	37
1983	736	3,793	24

Source: Rahim and Pennings (1987)

suitable and cannot be adopted without costly alterations. Since the number of potential users is constant or increasing very slowly, the growth curve is reaching its upper limit and the rate of growth is slowing down. This situation is further aggravated by widespread lack of computer workers, the high cost of computer labour and the lack of incentives on the part of the small commercial firms and other development-oriented enterprises to make substantial capital investments for computerization.

(Rahim and Pennings, 1987, p. 79)

Actors

The actors involved in the computerization process are of primary importance in their roles as change agents and experts. Actors are typically suppliers (vendors, consultants) but also academics and other policy makers. Due to the recent advent of computers in many sectors (e.g. industry) and also in many countries in the Third World, experience among users is often very limited or even non-existent. The computer user in a developing country is therefore forced to rely more or less entirely on different actors.

Actors' qualifications and familiarity with local information needs and practices are therefore of significant importance in order to provide appropriate and adequate computer solutions to users. But in many developing countries, real innovative thinking and new approaches to use computers in ways that are compatible with local conditions are rare compared to imitating ready-made solutions from the West.

There are different reasons for this. There is relatively little research among academics in developing countries on the development of, for example, alternative organization models appropriate for Southern conditions, which would enhance the understanding of and insight into local information needs. Local actors in many developing countries are therefore more familiar with the potential benefits of computers in the West than in their home countries.

Vendors, having the technical know-how to adopt computer applications, are either ignorant of local needs or unwilling to do so, due to business practices. IBM, as an example, follows very strict and formalized internal routines when judging whether a hardware or a software adaptation can be justified. A sales representative files a Request for Price Quotation (RPQ) regarding modification of a product in a sales situation. A favourable decision requires a strong cost/revenue factor due to the obvious uncertainty about whether sales will be high enough to cover the costs. As the limited potential in a developing country can not meet the sales volume required to keep the price low, the result is often that the request for adaptation is not accepted. As a matter of fact, with a priori knowledge among marketing personnel, it is likely that very few RPQs are filed.[4] A solution open to both suppliers and users is therefore to copy already existing solutions, which is often favoured by the suppliers.

This is very clearly reflected in most developing countries, where the dialogue between supplier and user, so important to the development of computer systems and applications in developed countries, is rudimentary. Products are available on an 'as-is' basis and with a great portion of responsibility left to the buyer/user to get his computer system to perform successfully.

Applications and users

Historically, computers were introduced into those sectors of the industrial society where formalized routines already existed and where the expected benefits of computerization were most adequately articulated. Organizations which followed these patterns were, for instance, banking/insurance companies and government institutions where electromechanical devices such as collators and sorting machines had long been used for the processing of high-volume data registered on punched cards.

A similar pattern can be observed as computers were and are being introduced into developing countries, with priorities given to transaction-oriented types of applications characterized by routine work, large volumes of data and relatively simple calculations. Such applications are, for instance, payroll, accounting routines and all kinds of statistics.

Computerization in developing countries

A recent report on computerization in a number of African countries shows the five major and most common application areas:[5]

Data and transaction processing systems are primarily devoted to the processing of large volumes of data that would take a long time to process manually. Statistical compilation activities such as population census, business and trade statistics; accounting activities like government expenditure and payroll.

Operational and management control systems are applications that are designed to improve operations, management control and decision-making capacities.

Sectorial information systems are database systems that generate information for planning and policy making in various sectors such as agriculture, education, industry, trade and transportation.

Multi-sectorial information systems are either an integration of various sectorial information systems or integrated planning and monitoring systems designed for policy formulation.

Planning and policy systems are econometric and mathematical models for resource allocation and optimization.

Table 2.5 Computer applications in five ASEAN countries (percentage of 1,846 firms having the applications)

Routine office data processing (billing, payroll, accounting etc.)	85
Business data analysis (forecasting, planning etc.)	28
Research (universities, industry)	14
Customer services (banks, railway tickets etc.)	12
Industry (production control)	11
Training (computer training)	9
Communication (data networks)	9
Data base (statistics etc.)	8
Design (computer-aided design, graphics)	4

Source: Asian Computer Yearbook, 1984

Data on 1,846 computer installations (77 per cent of total) in five ASEAN countries (Indonesia, Malaysia, Singapore, Thailand and the Philippines) showing the distribution of application areas is given in Table 2.5.

Technical as well as socio-economic constraints found in developing countries reduce the potential benefits of computers in on-line operation: poorly functioning telephone lines result in unreliable data transmission and insufficient infrastructure development turns out to be a serious obstacle when adopting a formal and systematic approach such as that of an advanced computer-based solution. The computer can not be integrated, as in a developed country with an advanced infrastructure, with other external functions and thus increase the overall performance, simply because the functional interfaces between the computer and those external functions have not been appropriately specified. Narasimhan gives an example of this:

> Software practices, such as distributed processing and word processing, assume new dimensions in countries where the telecommunication infrastructure is undeveloped, or where the local script and mode of writing differ radically from those in European countries.
>
> (Narasimhan, 1984, p. 10)

There are therefore only a few organizations/firms which have been able to move towards on-line oriented systems, where transactions are stored and retrieved in dialogue with a remote computer and with a simultaneous processing of data. Here, as before, we find banks and certain transnational companies as the principal users.

Users of computers in developing countries are mostly found in banks, insurance companies, government institutions (ministries, transport organizations, research centres, etc.) and the bigger enterprises, in more or less the same way as when computerization was first introduced in the developed countries, and for very much the same reasons.

The public sector is often the biggest computer user even if private enterprises are adopting computers at an accelerated pace. This is partly due to a relatively small private sector with an industry infrastructure still in its initial phase, but also to a not insignificant part played in public-sector computerization by military applications, which account for a considerable share of investment in computer technology.

Within the public sector, general authorities like electricity and water boards, central statistical offices, national banks, ministries of finance and national airlines normally represent the major computer users. Universities, although small in terms of computer capacity, are important users because of their ability to introduce the concepts of computer technology and methods to future decision makers. In those countries having a major national industrial sector, for instance oil, one finds relatively advanced computer applications.

Industrial computer applications

Computer applications developed for industrial use vary from the more administrative (e.g. production scheduling) to more technical types of applications (e.g. robot monitoring or automated warehousing). But as industrial activities also generate salaries to be paid and products to be sold we can classify the various computer applications in industry within a broad range, from the more administrative to the more process-oriented applications (see Figure 2.1). Examples of computer applications for the planning and control of production and material (production management) are given in Figure 2.2.

Production planning and control often focus on the utilization of resources needed for production, while materials planning and control focus on the material flow. In practice this distinction is difficult to follow strictly.

The scarce literature available on computerization in developing industries indicates that most applications are found in the upper half of Figure 2.1. There are, however, exceptions where relatively advanced applications occur, primarily in so-called turnkey installations where the computer is an integrated component for direct control of an industrial process, like in oil refineries and steel works.

Computers for production management with modules from Figure 2.2 are not common in developing countries. In manufactur-

Figure 2.1 Classification of industrial computer applications

production administrative applications
payroll, accounting, invoicing inventory management purchasing production planning material requirement planning computer-aided design computer-aided manufacturing production monitoring machine monitoring process control
production integrated applications

Figure 2.2 Some computer modules of a production management system

Direct production	Supporting functions
shop order release	master scheduling
shop order reporting	capacity planning
inventory control	purchasing
machine loading	cost calculation
plant maintenance	material requirements planning
quality control	product data monitoring

ing, very few computer installations seem to be used for applications below the inventory monitoring level in Figure 2.1. There are, however, exceptions showing indigenous and relatively advanced manufacturing applications for production scheduling and control as reported already at the beginning of the 1980s from India.[6]

It is, however, worth mentioning that different surveys from industrialized countries reveal a still relatively low penetration of computers for more advanced real-time processing. A study from 1983 is shown in Table 2.6.[7]

Table 2.6 Computerization in production management (distribution in percentages)

Function	Manual	Batch*	Real-time
Shop order processing	58	28	14
Master scheduling	55	42	3
Inventory control	30	53	17
Material reqirement planning	51	44	5
Shop order reporting	57	38	5

Note: *Batch is a non- real-time computer working mode

The low penetration of industrial computer applications in developing countries is often attributed to lack of capital and know-how, as well as to the attitudes of computer suppliers who find the market potential too small. Less discussed, however, are the more basic questions about the applicability of sophisticated planning methods in production environments that are affected by partly uncontrollable external factors, such as unpredictable material availability and constrained financial resources.

The problem of utilizing a computer-based system for material control in an economy that is characterized by a scarcity of exchangeable currency has been discussed by Bursche (1986). His conclusion from Poland is that the benefits of a material requirements planning system are severely reduced, since the requisition plans for material, issued by computer, have to be manually re-worked due to financial constraints for imported material.

A socio-cultural aspect of computerization

With increasing computerization in developed and developing countries the socio-cultural impact of computer technology manifests itself in a number of ways. 'The technology is here but the social awareness is lagging' was a headline in *Datamation* as early as 1979. And *Business India* pointed out that 'there is need to increase the awareness of people to the relatively new areas of technology such as microcomputers. Such an awareness is the necessary precondition of any social control over technology' (*Business India*, 1983).

Awareness can mean different things. For example, awareness about the computer language and its inability to interpret the meaning of metaphors and ambiguities that make every language rich and vivid, regardless of cultural belonging. (Wittgenstein used the word 'language game' to indicate the subjectivity as different social and ethnic groups develop their own rules for metaphors and meanings of words.) The computer can not cope with these language groups, designed as it is to work with a simplified language, stripped of all forms of ambiguities in order to fit into a logical Boolean algebra, a mental either-or world. This language is a direct result of a Western (European) scientific development, today largely integrated in the daily social life in the West.

One may ask what impact this specific Western language game may have on non-Western societies with different language games, as they adopt a Western way of formulating problems and articulating solutions. Kalman (1981) poses the question in the following way:

> It is generally admitted that information is not culturally neutral. By considering programming languages, it can be seen that they are largely based on the English language. Above all, current information is a reflection of a certain way of thinking and of a certain economic and social organization: it is the product of a rationalist and Western culture. Is this cultural dependence inevitable and must developed countries resign themselves to it?

This question opens up a new and fundamental set of questions concerning the cultural aspects of computerization. For instance, if technical processes are products of culture (rather than culture itself) as suggested by Malinowski (1931), is it then feasible to evaluate the impact of computers in different cultures without a thorough knowledge of local conditions? And do languages of different cul-

tures possess the same abilities (words, expressions) to express a rational computer language? (It is claimed that the Chinese language tends to counteract extreme formalization and abstraction in a refined logic. The tendency to argue and analyse phenomena in terms of a dialectical logic is reinforced in the language. Rigid 'A or not-A' categorizations are avoided [Elzinga and Jamison, 1981]).

Languages may thus be regarded as reflections of culture, and so are traditional rites and myths, being 'examples of mental activities of people and treated as communication in an unknown language' (Yalman, 1967). Considering this together with the notation made by Levi-Strauss, the anthropologist, that the structure of myths belongs to a different level of mental activity from that of language (Douglas, 1967), one is led to believe that 'software development' is far from merely a technical activity but comprises fundamental issues of a semantic as well as a socio-cultural nature. The computer language as a narrow subset of common language offers only a one-dimensional description of the many-dimensional faces of man.

Problems in computerization

From the previous section we find that the use of computers in many developing countries is characterized not only by low utilization but also by limited benefit to the user. In an attempt to discuss different types of problems facing computer users in the South the following three categories have been observed.

Computer users are exposed to *operational problems* due to technical constraints and lack of skilled personnel as well as other well documented reasons. These problems manifest themselves in low utilization of installed hardware but also in low productivity in very limited local software production.

An example of such operational problems occurred when the Indian government chose the British BBC computers for its computer-in- school project. The selected equipment never worked properly as it was not designed to operate in the high temperature and humidity prevailing in many Indian regions.

Computer users are moreover faced with *contextual problems*. Weak fit between models of Western design and applications in Southern contexts, semantic discrepancies in the wording and understanding of phenomena as well as references to different value systems and different concepts of rationality are all examples of problems that influence computer usage in developing countries.

Narasimhan gives an example of contextual problems. He first distinguishes between two kinds of application software: software products (packages) and software-supported systems. Software products, like accounting programs and other well defined solutions to well formalized problems, would still need to be modified before they become practically usable, but

> this is, in general, the simpler of the two cases to cope with. Software supported systems, on the other hand, almost always have to be tailor-made to suit the specific end-user environments. Even if similar systems are in use elsewhere, transporting them, modifying them, and fitting them to match local needs, may not be easy. In fact, in many cases, local specifications may be a preferred solution. The structures and expertise needed to create such systems may have to be dealt with on a case by case basis. Also transfer of knowledge in these system design areas is likely to be less straight- forward. But precisely these application areas are the ones of great immediate importance to developing countries. It is in meeting these needs in these application areas that available software production models in the developed countries are likely to be of very little relevance.
>
> Existing models in the developed countries are likely to be more valuable and applicable to applications related to increasing productivity in the industrial sector and to export in the software area. But even in these cases, the local conditions and contexts may require the creation and use of new structures. Industrial production establishments in the developing countries – even in the more advance ones – seldom have the level of information-awareness that is usually taken for granted in such establishments in the developed countries. Information processing practices, and software packages created to implement such practices in the developed countries cannot, therefore, be transported to developing countries and made to function effectively in a straightforward way.
>
> (Narasimhan, 1984, p. 8)

The third category of problems is *strategical problems* and refer to the diffusion of computer technology and how scarce resources, competing with other crucial demands in developing countries, are to be best utilized in the computerization process. Strategies also refer to the selection of appropriate applications with particular emphasis on developing problems such as basic needs, as well as the formulation of national policies for information.

Such a strategic issue is, for example, the relevance of the large-scale national information centres, which are prestigious institutions with very high computing capacity. The problem is that these centres have proved that they are unable to provide the specific and relevant information needed for economic development at local levels (villages, etc.).

Strategical problems can be regarded from a more general level in

a broader perspective on technology. Stewart suggests:

> Technology consists of a series of techniques. The technology available to a particular country is all those techniques it knows about (or may with not too much difficulty obtain knowledge about) and could acquire, while the technology in use is that subset of techniques it has acquired. It must be noted that the technology available to a country cannot be identified with all known techniques: on the one hand weak communication may mean that a particular country only knows about part of the total methods known to the world as a whole. This can be an important limitation on technological choice. On the other hand, methods may be known but they may not be available because no one is producing the machinery or other inputs required.
>
> (Stewart, 1978, p. 1)

She goes on to say that:

> If the technology in use is thought to be inappropriate, it may be inappropriate because world technology is inappropriate, or because an inappropriate subset is available to the country, or because an inappropriate selection is made, or for some combination of the three reasons. Confusion is caused by failing to distinguish between the three.

We have thus noted three groups of problems that confront computer users in the South. The three groups are very much interrelated and addressing a problem in one group can not be done without considering the other types of problems. In spite of this, most discussions seem to focus on problems within the first and the third group, i.e. operational and strategical problems.

The focus here is primarily on contextual problems and the purpose is to analyse the applicability of computer models in a Southern context. The applicability will be empirically studied through a case where the computer model, assuming a high degree of controllability in a predictable environment, is confronted with a reality influenced and determined by factors that are partly out of the organization's range of control, sometimes not even predictable. The applicability is tentatively assumed to be directly related to the degree of correspondence (or fit) between the computer model and the control structure of the studied company.

Before we proceed to describe and analyse the applicability of the computer model in a local environment, we need, however, to devote the next chapter to the question about models and their ability, or inability, to create relevant pictures of reality.

Chapter three

Model and reality – a conceptual discussion

Let us begin this chapter by making a very basic statement: computer programs are based on models of reality. A computer program is not a model in itself. This distinction is important to make here as we will be concerned with the two steps, from reality to model and from model to computer program, which literally determine the applicability and usefulness of computers. An example will illustrate this.[1]

An application area where computers have found extensive use is in accounting. Broadly, accounting is concerned with the collection of data relating to the activities of an enterprise and its use of resources, with the analysis of that data for decision-making purposes and with control over the use of the resources.

To perform his task the accountant needs a model: a simplified picture of the enterprise that highlights those specific issues that the accountant is interested in analysing, e.g. the cash flow. The model may also describe the pattern for decision-making as well as the control structure of the enterprise. The model is, of course, less complex than the enterprise itself in order to be manageable. It has therefore been deprived of most variables except for those considered most essential. For example, most models have been deprived of dynamism as they are to be used repetitively. Therefore, they reflect a static situation. Furthermore, accounting models, like many other models of economic or social realities, often represent different schools of thought. There may then be different, even competing models for the same reality, each model having its own view on cause/effect relations.

A computerized accounting system is built around a computer program residing on a computer with different peripherals as necessary (data storage units, printers, visual display units, etc.). The accounting system also has routines for data acquisition and data dissemination. The accounting program is based, i.e. designed and constructed, on one of the existing accounting models, using one of a variety of programming languages such as Cobol and Basic.

Already at this stage we may ask whether the model is appropriate. Is it too simplified, or too static, to be a relevant description of a particular situation. Also, does the computer language manage to interpret the model into an administrative tool in spite of its simplified false–true grammar? These are all questions to which we will frequently return in the following discussion.

The word model, the common denominator in the two steps above, has literally two meanings. One meaning is close to 'mapping or creation of an image of reality or part thereof'. The other meaning of the model is 'something worth imitating'.

For example, 'The Yugoslavian model' used to be regarded as an interesting socio-economic experiment between socialism and capitalism. And the Western model, as we briefly referred to in the introductory chapter, was a carrying idea in the development strategy in the 1950s and 1960s.

The failure to distinguish between the two meanings leads to confusion.[2] A computer program of an accounting model, for example, often referred to as a computer model, is regarded as something scientific, contextually neutral and therefore of higher value. As a matter of fact a computer 'model' is a transcription in a simplified language of an image that has been subjectively interpreted and created by somebody. Confusion may now arise if computer 'models' are discussed from the first perspective above, as something worthy of imitation, rather than from the second perspective where the question of usability and applicability of the theoretical models belong. It is not enough that a computer solution is *formally* correct, it must also be *practically* useful.

In the next two sections we will scrutinize the two steps, from reality to model and from model to computer programs, in more detail. In doing this we will show that the steps from reality to computer programs are much more controversial than is often realized. In the context of applicability of computers in developing countries, this discussion can not simply be ignored.

From reality to model

The interpretation of reality into a model is affected by three factors: what is the conception of reality, who has the conception and what is the purpose of the model. If there is a conception that reality is predictable then the purpose of the model may be to find just those variables that make prediction possible. But conceptions of reality differ among interpreters and the design of models are therefore influenced by the interpreter's own world view.

To make things worse, our conception of reality may be based on other images, or models, that we create or which are created for us by others. It is therefore a tendency, if seen in a wide time perspective, that certain models and conceptions of reality amplify each other into solid basic models. Such a basic model appears in the Western view on certainty and predictability, which will be discussed below. The relationship between the model and our conception of reality means that adoption of a model is also to adopt a world view or a conception of reality.

We will briefly describe how changes in the conceptions of reality in the West (or rather in Europe) has led to a world view that is predominant in the West and which has shaped the concepts of rationality and time into their specific Western forms. We will also contrast this model to Southern contexts and, based on recent theories and ideas, question some of the fundamental principles of this Western certainty model.

Model and reality – a time perspective

The life of primitive man has always been a struggle with Nature. Food, shelter and protection against harsh climate and wild beasts were his basic needs. If Nature could be conciliated, man would benefit. From this point the concept of rationality has developed.

In Ancient Greece this rationality was based on a philosophy where the universe was ruled by lawful order. Man, being part of the universe, therefore had to act in accordance with these rules. In the European Middle Ages, Nature lost its positive influence on man who instead oriented his intellectual mind towards the divine.

During the Renaissance in Europe (approximately from the fourteenth to the beginning of the sixteenth century) that marked the transition from the medieval world to the modern, man began to fight prejudice and superstition that had been prevailing in the Middle Ages. The relationship between Man and Nature was rediscovered but unlike the Ancient Greek ideal, man considered himself superior to Nature which should be tempered. Nature's resources should be exploited and used in the service of Man. As the new era dawned, science became an important tool.

At this time we can see the first sign of today's rationality concept emerge. (Still, however, the very basic notions of rationality remains: that Man will benefit if Nature is conciliated.) Many philosophers and natural scientists appeared and unified within this rationality concept. 'Knowledge is power', said Francis Bacon (1561–1626) and suggested that knowledge about the causes of natural events is fundamental to the prediction of future events and, through this, to master Nature. This knowledge of the causes could be found by making experiments. To experiment meant to study the course of events under simplified and controllable conditions. This experimentation as an intellectual approach is very typical of Western man and has dominated scientific tradition ever since. As we shall see, signs are emerging that begin to question this scientific ideal.

Different conceptions of reality gave rise to different models of, for example, how to predict the future. The Delphi Oracle, in Ancient Greece, could predict future events by reading signs in nature. In other cultures, looking into viscera opened windows to the future.

With the experimentalists the world around us, although complex and not wholly understandable, could be interpreted in simple models as a kind of intellectual tool with which life could be not only described and explained, but also formed. Such a model was, for example, the laws of planetary motion. Newton (1642–1727) proposed that by knowing an initial condition and the physical laws that determine how states change, any other future state could be predicted. The remarkable achievements in making predictions encouraged European natural scientists to proceed with their experiments and to build models. In the late nineteenth century, this experimenting and model building was also extended into the social sciences.

As we have already mentioned, there is a strong relationship between the world view that we have, the knowledge we have acquired through models and the rationality concept we are using. In the dominant European scientific world view of the late nineteenth century, often referred to as the Vienna school and logical empiricism, the criteria for what was rational, what constituted knowledge and what should be considered as good science were regarded as unbounded by culture and independent of time.

With Thomas Kuhn[3] this view on science was, however, criticized and there is now a growing awareness that scientific knowledge is dependent both on its cultural context and on time. It is therefore not possible to formulate criteria for rationality and scientific knowledge and, hence, for the construction of models that are independent of cultural and historical contexts.

For this reason it is unfortunate that the Western rationality concept and related models have been so uncritically adopted by other cultures. Scientific rationality in India is thus synonymous with Western science whereas non-Western forms of science are regarded as irrational. We can probably find the background to this from European colonialism, particularly as similar patterns are found in many other developing countries. Models of Western origin therefore reflect a world view that is deeply rooted in the Western, or rather

European, way of thinking. This world view can be characterized as analytic and intellectual rather than perceptive and intuitional.

The question whether the intellectual mind or intuition is best aimed at understanding and creating knowledge about the world around us has been a traditional subject among philosophers. Hume (1711–1776) raised questions about causality that are still relevant today and Kant (1724–1804) accepted that certain things around us can only be grasped through the intuitional mind. Bergson (1859–1941), in turning against the mechanical interpretation of life, even placed intuition before the intellectual mind, as a quality to understand the world around us.

Today, on a more pragmatic level, we can see signs in the social sciences (e.g. organization theory, business studies) that indicate a growing awareness about intuition as an important quality among, for example, business managers. This is partly a reaction to the grand visions of the 1960s and 1970s with the constructions of very sophisticated models of management information systems. Let us therefore briefly give some reasons why these, and similar, models did not work as expected.

Model and reality – some critical points

Models are always based on somebody's interpretation of reality. This incontrovertible fact leads us to raise questions regarding the objectivity of models. For example, as the model is partly a reflection of the originator's own world view and perception of problems and solutions, what is the likelihood that it is applicable in cultures with other value patterns and life styles than those of the originator? How do we know that a model, being a simplification of reality, retains just those relevant parameters that describe the cause-effect relationship to be studied? How do we know that the isolated parameters are sufficient to describe or group the problem? And how would we know how many parameters are needed? Hofstadter illustrates the problem as follows:

> There must be many 'just plain' rules. There must be 'meta-rules' to modify the 'just plain' rules; then 'meta-meta' rules to modify the 'meta-rules'; and so on. The flexibility of intelligence comes from the enormous numbers of different rules and level of rules. The reason that so many rules on so many different levels must exist is that in life, a creature is faced with millions of situations of completely different types. In some situations, there are stereotyped responses which require 'just plain' rules. Some situations are a mixture of stereotyped situations – thus they require rules

for deciding which of the 'just plain' rules to apply. Some situations cannot be classified – thus there must exist rules for inventing new rules ... and so on.

(Hofstadter, 1980)

The kind of difficulty in the interpretation of reality into a model that we have tried to describe here could be termed subjectivity in interpretation. Another difficulty encountered is the problem of linearity assumptions.

In the discussion of model and reality in a time perspective we could refer to the Newtonian world view as the basis for today's conception of predictability and certainty and also of rationality, all from a Western perspective. In this Newtonian world view predictability of the future is gained by knowing the initial state and the natural or physical laws that transform a state into another. The remarkable results achieved by Keppler (1571–1630) and others in applying Newton's theories to predict planetary motions encouraged scientists throughout the following centuries to apply similar theories to a great variety of problems of technical, economical and even social nature. The mathematical tools that were developed and became more and more sophisticated were, however, based on one very fundamental assumption, namely that the relationship between variables in the models was a linear one. The meaning of this is that the sum of any two solutions to a dynamic problem, at different points in time, is in itself a solution.

Recent observations show, however, that structurally simple systems can be dynamically (i.e. time-dependent) very complex. The idealized picture of dynamic systems as something well defined and stable, as in Newton's world, is not generally valid. The explanation of this is that the assumptions about linearity is a very serious restriction that can mostly not be justified. This applies not only to research and natural science but also in the everyday life of politics and economies.

As a matter of fact, we can not take for granted that natural systems are even in principle predictable. Weather systems are good examples. In the deterministic, mathematical expressions that describe the future of the weather, solutions are hidden that exhibit complete chaos, something that also correctly describes the weather. That such utter chaos can emerge from deterministic equations is one of the remarkable discoveries of modern mathematical physics. Further examples of systems that we must assume to be in principle not predictable are systems describing the stock market or the international economy.[4]

The third difficulty in the interpretation of reality to model has to do with uncertainty. The interpretation itself is, as we have seen, exposed to uncertainty as to the solution of model parameters. But uncertainty may also appear in reality itself, which was observed in the quantum theory. If uncertainty is accepted as a characterization of reality, this must be reflected in the model. Recent experience in organization theory has, for example, resulted in less ambitious models for management and control. Instead, models are developed which can cope with uncertainty and contingencies in the world around us.

A recent statement in *The Economist* (August 1989) will serve as the conclusion of this section: 'The triumph of the twentieth century is that it has purged itself of certainty'.

From model to computer programs

The step from model to computer program means, as the step from reality to model, a further simplification. In the first step this was deliberate because the model should be a simplified, yet trustworthy, image of reality. In this second step, however, the transformation becomes a simplification because the computer technology sets limits. This simplification is therefore not deliberate.

A major limit is of course the computer program itself that must be designed in accordance with a pre-defined logical structure to fit with the technological constraints. Ambiguity in the model, implicit or unavoidable if the model is to reflect a social system with human integration, is not allowed in the computer program, simply because the computer program can not by itself judge whether a conclusion arrived at in the program is reasonable or not. Every situation that may appear in reality must therefore be a priori foreseen in the model to be reflected in the program, otherwise it will not be recognized at all. It is thus a simplification and also a limitation in the usability of the model, that all forms of ambiguity must be avoided in the computer program.

Models help us to understand the world around us and to communicate this understanding to others via text-books, discussions, articles, etc. This communication is based on language, albeit the language is sometimes inadequate to express specific phenomena and subtle meanings. As long as the communication takes place in a cultural domain with a common frame of reference for abstract reasoning and metaphors, these difficulties can be overcome. But as soon as this common reference frame can no longer be assumed as a basis (for example in communication between different cultural domains with different value patterns and intellectual traditions), communication may become distorted and meanings become defective or misleading.

In this respect, computer language can be regarded as an expression of a specific culture where the language is based on formal logic and deprived of inexactness and ambiguities. In this culture, exact wording in the formal language and semantic accuracy are cornerstones. We can therefore say that the computer language is too straightforward to serve as an interpreter of the conceptual thinking behind many models.

Models can be regarded as a kind of intellectual tool that help to understand the world around us. One and the same reality can be interpreted by different models, each model having its perspective and highlighting particular phenomena. But one and the same reality can also be interpreted differently by different interpreters and thus produce different models. Such models often represent different schools of thought and show thereby the inexactness and uncertainty in our attempts to describe reality. General models are therefore no more general than the underlying world view of their originators. In other words, no model can claim absolute credibility. 'Our knowledge is never completely certain' said Karl Popper, one of the most influential philosophers in the theory of science.

But when a model is transformed into a computer program it has a tendency to become a generalization. There are primarily two reasons behind this. First, a computer program is regarded by many as the result of a consensus decision where the most appropriate model has been selected to be transformed into a computer program. The second reason is that those suppliers who invest in producing this particular computer program promote their product as being as generally applicable as possible in order to maximize the volume of sales.

As long as the model can be traced back to the originator, validation of the model can be done by evaluating the arguments of its originator. In the computer program the bridges back to the underlying model and its originator are in general blurred. The transfer from model to computer programs is therefore also a step of 'impersonalization'. The view of a computer program as an expression of formalized logic has contributed to the conception of a higher veracity. This will again give us the reason to take a brief look back in time to see how the ideas of formal logic developed into computer programs.

Model and computer program – a time perspective

In the middle of the seventeenth century, as the scientific development towards great accuracy in measurement created needs for artificial modes of expression, formal mathematics developed. Leibniz (1645–1716), one of the first to use an axiomatic structure to describe thinking, based his theories on a formal method where sentences could be described in a formal language and conclusions could be drawn, as in solving an equation.

A century later, Boole (1815–1864) presented his logical algebra with a further systematizing of the analogy between thinking and algebraic operations. Boolean algebra was first used in set theory and in probability theory and not until the end of the 1930s was the most important application area realized: mechanical computation. It was Shannon who unified the two development paths (formal logic and mechanical computation) by proving that what happened when electrical components (e.g. transistors) were connected could be interpreted by Boolean algebra. From then on the development into today's computer technology took place.

This formal logic is thus to a great extent realized in today's computer technology and, perhaps, even more in the systematic approach to problem solving (i.e. flow-charts), which is also the structure of computer programs.

Conclusions

In this chapter we have attempted to identify a number of reasons why Western computer 'models' are not a priori applicable in non-Western contexts.

It is now the intention to seek confirmation that there are areas where there is a misfit between model and reality. We will do this by analysing computerization in an industrial enterprise. But before we begin this analysis it will be necessary to find out what typically characterizes a Western view, or model, of industrial production.

Chapter four

Industrial production as a model

In this chapter we will narrow the general discussion of models from the previous chapter and restrict ourselves to models of industrial production. There are several reasons for this choice. The first is that industrial production in developed and developing countries follows the same organizational model. The second is that computer systems for industrial applications are assumed to be applicable generally. The third reason is that industrial production is influenced by its environment, which affords an entry to the question of applicability of computer programs in different contexts.

Generalization of organization models

Major industries in Third World countries have an organizational form that very closely follows a traditional Western pattern of organizing and managing industrial production. In this Western view lie assumptions about production problems and how planning and control can reduce or eliminate these problems. The assumptions have been further elaborated and ideas have been formalized into computer programs for production management.

The formal resemblance between a Third World industry and the traditional Western production models has therefore brought about an illusionary expectation that the computer systems should be particularly applicable in those industries. But as production is linked to the outside world through the flow of materials, a good deal of organizational behaviour can be understood only by knowing something about the environment the organization is in and the problem it creates in obtaining resources. In a non-Western environment, that is characterized by constraints and contingencies partly not found in the West, we must therefore ask what is the guarantee that a Western model of production, and hence also the computer programs, will fit?

This question goes back to the organization research of the 1950s and 1960s and a number of studies showing how different organizations partly exhibit very different characteristics. An important finding in these studies was that an organizational structure or a planning system or a particular management style that works well in one organization may be inappropriate in another. The reason is that the organization can not be regarded as isolated from its environment. Different environments with different characteristics therefore lead to different organizational behaviour.

With this newly-won insight organization theory abandoned its former ambition to create a general model for the study of organizations. Instead it has been more fully accepted that it is only by studying an organization in its environment that its behaviour can be understood. This theoretical approach is often referred to as the 'contingency theory'.

As this theory is less formalized (and less possible to formalize) than earlier organization theories, it has not produced any simplified organization model. For this reason most computer programs still reflect the earlier organization models and ideas of earlier theories, even if their explanation value has turned out to be relatively limited.

The methodology we will use to assess the degree of fit between model and reality and, hence also the applicability of the computer programs, is to use the contingency theory referred to above. But first we will give some details of the traditional Western view of production.

A traditional model of production

Industrial production can be described in different ways, depending on what is to be highlighted. In this study, with the focus on material and how material needs are planned and co-ordinated, it has been fruitful to conceive production as a flow of raw materials and components transformed into products. As material is used in the different production steps along the line, information is collected and disseminated as appropriate to suppliers for subsequent material supply. The material/information flow can be regarded as a flow of energy, used in a broad sense to include both material and immaterial quantities.

The planning and co-ordination of activities as material is transformed into products involves the interaction of many different departments and functions. A schematic overview of this interaction from a production management perspective is shown in Figure 4.1. Sales and forecasts together constitute the basis for a production (assembly) programme from which the requirement for parts, components and sub-assemblies (material) is determined. The pro-

The contingency theory

The focus on contextual problems in the use of computers implies that the interaction between organization and environment as well as the impact exerted by the environment on the organization becomes vital. This leads to the search for a theoretical approach which can handle categories identified as significant such as interdependency and uncertainty.

Theory and model

In 1956, James Thompson called on readers of *Administrative Science Quarterly* to contribute to the building of an applied administrative science applicable not only to Western societies but also in other parts of the world (Thompson, 1956). Thompson's motivation for this step was a dissatisfaction with existing administrative theories and models, which he found appropriate to one type of cultural setting only.

In a paper published after his death, Thompson (1974) returned to this theme by drawing on his earlier developed concept about organizations as open systems where terms like uncertainty, interdependence and rational behaviour constitute cornerstones of the concept (Thompson, 1967). As this concept seems to be well suited for our purpose it has been adopted as a theoretical framework. This choice of theoretical framework is naturally not indisputable, and at a later stage (Chapter 11) the adequacy of this choice will be discussed.

The conception of rationality as used by Thompson (but also by Katz and Kahn [1966] and by March and Simon [1958]) breaks with a traditional view of organizational behaviour (e.g. the simplified rationality of 'the economic man') and accepts the 'cognitive limits of rationality' (March and Simon) or 'bounded rationality' (Thompson) where:

choice is always exercised with respect to a limited, approximate, simplified 'model' of the real situation, the chooser's definition of the situation.

(March and Simon, 1958, p. 130)

But rationality is a contextually related concept, and assumptions about rationality are facilitated if the attempts to reduce uncertainty can rely on abundant resources (money, skills, materials, etc.). As this form of abundance is not a privilege in Third World countries, we will find several reasons in the following discussion to return to the concept of rationality.

The theory

As a point of departure we will take Thompson's (1967) notion that complex organizations can be regarded as open systems, faced with indetermination and uncertainty but at the same time requiring determination and certainty, since their behaviour is subject to criteria of rationality. Adopting Thompson's view implies that we accept a company as a system, i.e. an organized group of components (subsystems) linked together according to a plan to achieve a specific objective (Awad, 1979).

Thompson regards uncertainty as the basis for organizations to behave as open systems; rational organizational behaviour, in Thompson's view, is to close the system, save for a limited number of gate-ways ('boundary functions') which are designed to handle uncertainty in such a way that its influence is reduced or eliminated.

Thompson distinguishes various kinds of uncertainty to which organizations can be exposed. Generalized uncertainty, with no or only vague patterns of cause/effect understanding, does not give a firm enough basis for recognizing alternatives or establishing control patterns. When cause/effect understanding is present, uncertainty stems from interdependence with an environment that may be uncooperative. Under this contingency the organization tries to reduce uncertainty through regulation of transactions at the boundary. When boundaries become regulated and there is cause/effect understanding, uncertainty remains in the internal interdependence of components which the organization seeks to control by co-ordination. It should be added that an uncooperative but stable environment may constitute a framework for the organization that is not uncertain but rather constrained.

Rational behaviour presents itself in different ways. One way is that organizations seek to seal off their core technologies from environmental influence, where core technology denotes that part of the organization where the main activity is performed, e.g. manufacturing and assembly.

The behaviour of this technical core is, however, always an incomplete representation of what the organization must do to accomplish desired results:

> Technical rationality is a necessary component but never alone sufficient to provide organizational rationality, which involves acquiring the inputs which are taken for granted by the technology, and dispensing outputs which again are outside the scope of the core technology.
>
> (Thompson, 1967, p. 19)

In the following we will be concerned with two of the three major activities of the technical core: input activities (material acquisition) and technological activities (manufacturing and assembly), leaving the third activity (marketing and sales) outside the scope of the study. It should be noted here that it is the activity of marketing and sales that is left out, not the market and its influence on production.

The empirical studies discussed in the following chapters show that, for the actual Egyptian company examined, there is interdependence within the organization between activities, as well as between the organization and its environment, and the flows of material and information must be co-ordinated both in the exchange with the environment and as they pass through the organizational structure.

Interdependence between the organization and its environment means, as suggested by Thompson, that rational organizational behaviour is based on the assumption that the organization can be described as an open system. In this open-system approach the rational organizational behaviour is subjected to environmental constraints which the organization must accept, a *state d'état* which is predictable from the organization's point of view, at least within the time period considered in this study. The rational organizational behaviour is, furthermore, met with environmental contingency that may vary in an unpredictable way, i.e. not controllable by the organization. Finally, rational organizational behaviour, in sorting out those variables that are not controllable from within the organization, resorts to the closed-system structure for its technical core where internal interdependence, assured of predictability, can be co-ordinated.

The environment here is the material and equipment market in Egypt and abroad, part of the Egyptian infrastructure such as transportation and various finance and government institutions. The principal feature of this environment, from the point of view of the study, is that basic production inputs such as vital components and equipment can not be supplied by the local market but must be imported for scarce financial resources that compete with other urgent demands of the country.

Environment systems can now be classified so that the more heterogeneous they are the more constraints they present to the organization, and the more dynamic they are the more contingency the organization has to face. A company may be exposed to influence from the environment in both ways so that, for instance, the acquisition of material and production equipment is constrained by the country's scarcity of convertible currency and contingencies make the prediction of lead times for material supply difficult.

37

For interdependence between the organization and its environment the rational way to cope with environmental uncertainty under the assumption of cause/effect understanding, can be to use the boundary system as a buffer for external influence. An example is the ordering of material by the purchase department that negotiates with suppliers and other external parties involved. Rational behaviour also implies, unless environmental influence can not be successfully buffered, that the core technology, represented by a company's production functions, adapts itself to reduce the impact of uncertainty. Such adaptation may be the deliberate extension of lead times through earlier release of shop-orders, thereby having more active shop-orders to select from, if and when re-routing of orders is called for due to machine failures or material shortage.

The organization tends to structure itself in terms of open and closed sub-systems so as to cope with uncertainties and to eliminate their effects. The technical core of the organization, here associated with a company's production sub-system, tries to close itself from external influence in order to concentrate on the task of co-ordination of internal activities, thereby reducing uncertainty that stems from internal interdependency. For this internal uncertainty Thompson (1967) has identified three types of interdependence, and hypothesizes that co-ordination depends on these three types.

The most simple type is pooled interdependence where departments affect each other only to the extent that they share the same pool of resources or affect a common constraint in an additive manner, for instance in the use of funds for the acquisition of material and spare parts when the financial supply is limited. The second type is sequential interdependence where the output of one department is the input to another, e.g. manufacturing feeding assembly as in Figure 4.1. Finally there is reciprocal interdependence if two departments input as well as output to each other. Follow-up and control of shop floor activities from the planning and control department is an example of reciprocal interdependence. It should be noted that reciprocal interdependence also implies sequential and pooled interdependence as well.

For internal interdependence Thompson now suggests three different types of co-ordination that correspond with the three types of interdependence. The first type may be achieved by standardization which:

> involves the establishment of routines or rules which constrain action of each unit or position into paths consistent with those taken by others in the interdependent relationship. An important assumption in co-ordination by standardization is that the set of

rules be internally consistent and this requires that the situations to which they apply be relatively stable, repetitive and few enough to permit matching of situations with appropriate rules.

(Thompson, 1967, p. 56)

The second type is co-ordination by plan that:

involves the establishment of schedules for the interdependent units by which their actions may be governed.

(Thompson, 1967, p. 56)

The planning process involves the establishment of schedules and targets to govern the actions of the interdependent units. Planning is appropriate to more dynamic situations than is standardization. Plans quite often consist of changing the parameters in established decision rules.

Finally, the third type of co-ordination is mutual adjustment or 'co-ordination by feedback' that:

involves the transmission of new information during the process of action.

(Thompson, 1967, p. 56)

The task is sufficiently unpredictable that pre-established decision rules and plans cannot be prepared. Thus as the predictability of the task changes, the appropriate form of co-ordination changes from standardization to planning to mutual adjustment.

(Thompson, 1967, p. 118)

Thompson now makes two important observations about co-ordination and interdependence.

A. With pooled interdependence, co-ordination by standardization is appropriate; with sequential interdependence, co-ordination by plan is appropriate and with reciprocal interdependence, co-ordination by mutual adjustment is appropriate.

B. The three types of co-ordination place increasingly heavy burdens on communication and decision. Standardization requires less frequent decisions and a smaller volume of communication during a specific period of operations than does planning, and planning calls for less decision and communication than does mutual adjustment.

Figure 4.3 summarizes the different types of uncertainty we have identified and what organizational behaviour, according to Thompson (1967), can be expected under norms of rationality.

Figure 4.3 Source of uncertainty and organizational behaviour

Uncertainty due to	Organizational behaviour
no cause/effect understanding (generalized uncertainty)	*ad hoc*
interdependence with the environment	e.g. boundary functions
interdependence within the organization	co-ordination

Source: Thompson 1967

Co-ordination of activities to master uncertainty thus requires information, and the amount of information needed depends on how much uncertainty there is before the activities are performed (J. Galbraith, 1970). Not only the amount of information but also reaction time (or 'frequency of decision' in Thompson's words) characterize this type of co-ordination. Feedback is thus present in all three types, albeit with different time frames.

With reference to Ashby (1956), there is an upper limit in the utilization of information to reduce uncertainty, determined by the information-handling capacity of the closed sub-system.[1] This upper limit is seldom conceived of in practice compared to the more widespread opinion that 'the more information the better decisions' for production management (Plossl, 1973, Smolik, 1983).

The approach to an organizational behaviour thus presented, based on the open/closed system analogy, constitutes the framework to be used for the coming company analysis.

Production planning and control

Planning and control of production can be regarded as a way of coping with uncertainty, in Thompson's (1967) meaning, and we have already, in the previous section, referred to uncertainty in the different markets that surround and exert influence on production. There is also uncertainty in the socio-technical system of the organization: in the performance of equipment, in the skills of employees, in the relevance of information, etc.

The base for production is the products. This means that the way a product is put together from different parts, the material requirements for the product and the equipment needed for its making are all fundamentals of the production process and define, in principle, the planning requirements. Uncertainty in one or more of these fundamental issues, for example where a specific part can be acquired or what parts are included in a certain product, contributes to lower performance in production.

From Figure 4.2 we note how production is interrelated with external markets, e.g. the resource market, and it lies within the scope of production planning and control to eliminate the influence from uncertainty in this market. Planning and control of production therefore means to interact with the boundary functions, e.g. the purchase and sales departments as indicated in Figure 4.1.

As all concerned departments need, for example, basic product data, there is pooled interdependence between departments. There is also a standardized way to co-ordinate so that this basic resource is available to all concerned departments, for example through part numbers by which each item in the total set of thousands of different items can be individually referred to by designers, purchasers, tool makers, etc.

Planning of production assumes that the production goals are set, that items are adequately specified so that the right material can be acquired beforehand, and that sufficient production capacity is available. Difficulties in planning arise due to a number of factors such as uncertainty in markets or to uncertainty in production capacity or simply because goals are not realistic. For co-ordination by production planning it is therefore imperative that planning parameters such as acquisition lead times are correct. Due to the large number of planning parameters and to the necessity of a uniform handling of data, standardized routines for the creation and maintenance of this data is imperative. Planning results in planned production times, equal to the sum of standard times. As plans are executed, actual performance is compared to what was planned.

Uncertainty that is not eliminated through co-ordination by planning may require specific measures as events occur which violate the plans. Such measures may be predefined and the most adequate measure be selected, for example re-routeing of shop orders in case of sudden machine failure. For events which can not be predicted, such as an unplanned rush order, co-ordination may be based on *ad hoc* measures such as negotiations between departments as a means to reduce uncertainty.

Production planning and control can thus be seen as a way to reduce uncertainty. However, as uncertainty takes different shapes in different contexts, the tasks for production planning and control may vary. In the West, where the resource market as well as the finance market are relatively stable, the product market is more unpredictable. Production planning is therefore geared to flexibility and to adaptation of production to meet changes in demand. In many Third

World countries where the resource market and the finance market are the uncertain ones, production planning is primarily the planning of material acquisition and supply.

In Western production models, the core of the materials control problem is commonly described as a cost optimization problem whereas external conditions (suppliers, transport) seldom present any major hindrance to fulfilling work orders or purchase order quantities. For most Third World industries, however, materials supply is perhaps the most basic problem of production.

Computerized production control

Computer applications for use in industry have been developed for production planning, for robot monitoring, for automated warehousing, etc. But as industrial activities also generate salaries to be paid and products to be sold we can classify the various computer applications in industry within a broad range, from applications for the administration of production to production integrated applications. We will again refer to Figure 2.1. In the figure we classify industrial computer applications in a scale from more administrative to more production-integrated. This classification does not, however, reflect the complexity of the applications. We will therefore discuss the qualitative difference between two of the applications of Figure 4.4, inventory management and production planning.

Two industrial applications

What the two applications have in common, with each other and with

Figure 4.4 Classification of industrial computer applications

production administrative applications
payroll, accounting, invoicing
inventory management
purchasing
production planning
material requirements planning
computer-aided design
computer-aided manufacturing
production monitoring
machine monitoring
process control
production integrated applications

the rest of the applications of Figure 4.4 is that reality is mapped on a model, and the questions we may ask are how well does the model reflect reality, and how well should the model reflect reality in order to be an adequate representation. This can be seen as a static problem. The dynamic nature of the problem arises when the model is fed with information and becomes the model in action. This will be the aspect of the following comparison.

The inventory management application

The kind of reality we wish to represent by the model is the following. In a production company materials (raw-material, components, semi-finished goods) are stored before use in production. The number of items may vary but can always be calculated. The model is constructed to include parameters such as item number, store position, stored quantity. Characteristic for this model is the a priori knowledge of how the model must be designed and the exact content of the information set. (It is, for example, irrelevant to know item price.) The application is completely deterministic in the sense that all relevant transactions can be performed with the model and its well defined set of information. Alternatively, we can tell exactly what transactions cannot be performed by the model.

This model may seem trivial but is representative of a great many computer applications. We will in the following refer to them as 'keeping-track-of' applications.

A production planning application

Reality in this example is represented by a company where planning of production is based on parameters such as market demand, actual and expected, supply of material and available production facilities. The objectives of the model are to co-ordinate these parameters and to present a production plan for future time periods.

The nature of this problem differs in a significant way from the previous one: it cannot be determined a priori whether the model and/or the available information set is sufficient to present an acceptable solution. We do not know what information would be required to obtain a better solution. And besides, what would be a better solution? The problem of indeterminism in this example appears as we regard the organization as an open system, influenced by an unpredictable environment.

Thompson's (1967) statement that 'whenever the organization behaves as a closed system, conventional theory applies' will therefore serve as a point of departure for the following discussion where 'conventional theory' will be synonymous with a computerized production management system. We will therefore, in the last section of this chapter, present the design of such a computerized system which will serve as a representative for similar systems. The system is IBM's COPICS.

COPICS fits well with the view that regards production as a closed system where uncertainty primarily stems from interdependence between functions and departments within the organization. It is well adapted to deal with fluctuations in the production flow and to make adjustments, provided there are well-defined and articulated standards on which the various control algorithms can be based.

COPICS – a production control system

COPICS, an acronym for a Communication Oriented Production Information and Control System, has for a long time been IBM's major application concept for production industries. It is a modular concept, each module being an independent application package (e.g. Stores Control) where all packages are related to each other through a number of data bases. The whole concept is based on and is a realization of an industry model, described thoroughly and in great detail in a set of documents (COPICS, Volumes I–VII). COPICS is conceptually a further development of earlier IBM products.

Some development steps towards COPICS

In the early 1960s IBM released two software products to be used as standard software in material handling and production contexts. IMPACT, an inventory control system used in retail and manufacturing industry, and BOM, a bill of material system. These two products were followed in the 1960s by a capacity loading and sequencing system (CLASS), which was an improved version of a German shop loading system (KRAUSS). CLASS was subsequently further improved to CAPOSS (Capacity Planning and Ordering Sequencing System) and to a further extended version, CAPOSS-E, still a current IBM product for this purpose.

In the middle of the 1960s, the first modular concept for production control took form as a product. PICS – Production Information and Control System – was the group name for six modules, some of them already existing software products (BOM, CLASS), but also new modules, e.g. MINCOS, a material and inventory control system that also contained an algorithm for material requirements calculation. With PICS, the first attempt was made to define an integrated production control database.

In the early 1970s, as on-line and real-time facilities developed that enabled direct communication between user and computer, PICS was enhanced and transformed into COPICS, a communication oriented version of PICS. COPICS currently has twelve modules, all grouped around a set of interrelated data bases.

COPICS, like other similar system concepts with the objectives of supporting production planning functions in industry, has in its design and approach adopted the most recent advances in computer technology. Fast data communication between the user and the computer, capability of storing very large data volumes and a variety of possibilities to present data in the most convenient form are only a few development features that are made use of in COPICS. Conceptually, however, COPICS outlines a conventional approach for a conventional industrial setting that dates back to the 1960s. The enhancements and new developments comprise additional modules of conventional character rather than more innovative thinking within the production control concept.

A COPICS overview

COPICS is a family of concepts that outlines an approach to an integrated computer-based production control system. The concepts are thoroughly documented and presented in seven volumes each containing twelve chapters, each chapter addressing a specific function (e.g. shop-floor management). Certain of these COPICS concepts have been implemented by IBM into a number of separate software products, the COPICS implementation products. These products or application modules are shown in Table 4.1.

Table 4.1 The COPICS modules

Customer order servicing
Inventory planning and forecasting
Inventory accounting
Advanced function MRP
Bill of material
Product cost calculation
Purchasing
Receiving
Shop order release
Online routeing
Shop order load analysis and reporting
Facilities data management
Plant monitoring and control

Source: IBM *COPICS: Manufacturing Application Overview*, 1983

Each module can be used independently but requires access to data bases which are also used by other modules. As data created as output in one module may be needed as input in another module, many modules are in practice closely dependent on each other. The very complex and highly integrated logic structures within and between modules makes it difficult for the user to decide which modules need or need not be present in a specific configuration for satisfactory performance. For this purpose IBM provides optional utility software (Environment Table) as a means of communicating to all installed COPICS implementation modules details about which other modules and data bases are installed.

COPICS can be characterized as a 'net-change oriented' concept, thereby implying that variations are taken care of as they occur and are continually compared to standards and reference values which have been submitted to the system. Fast reaction to events is implemented through real-time processing of data and from the use of action queues from which appropriate, pre-defined actions are issued in response to occurrences. In our terminology we may state that COPICS is aimed at co-ordination by mutual adjustment.

COPICS presupposes an organizational behaviour where decision structures are well defined, where the needs of information can be well articulated and where profitable production is a well-defined concept. It assumes a functional rationality in the sense that lead times are always possible to determine or at least estimate and that a plan is a manifestation of the will rather than a formal document only.

A description of COPICS is a description on three different levels. The basic model, defining the scenario of an industrial environment, the COPICS pattern company, is presented in the COPICS concept volumes. On this description level all assumptions are made about what characterizes the 'COPICS world'. These assumptions are then, explicitly or implicitly, conveyed into the COPICS model and then, in turn, into the implementation programs. A description of COPICS may therefore also be a description of the model or of the implementation programs.

Figure 4.5 The COPICS description levels

The COPICS pattern company	B0
The COPICS model of the pattern company	B1
The COPICS implementation programs	B2

An example will illustrate the differences between the levels. Inventory accounting is faced with the problem of keeping inventory records up to date, notwithstanding unauthorized picking or missing delivery notes, i.e. a scenario of a typical industrial environment (B0). In the COPICS image of this reality (the COPICS model, B1), the situation is recognized and it is admitted that no formal system will prove efficient unless strict control over the reporting of inventory transactions be established. It is implicitly assumed that the organizational behaviour includes control procedures that can guarantee full reporting. Therefore, in the program description on level B2, with awareness of all ruling preconditions it is simply stated that the program is capable of displaying all inventory status information for an item in summary or by location.

On a program level, which is normally the only level where general application software may be studied and evaluated, there is no explicit description of a more conceptual nature, i.e. of the industrial scenario from which the programs have been designed and developed.

What makes COPICS different from other application packages for production control, and therefore also more open for a conceptual analysis, is the availability of the complete description of the COPICS model, explicitly stated as follows:

> The basis for COPICS is a well documented conceptual systems description that frames the development of an integrated database oriented material- and production-control system.

In evaluating the applicability of computer programs for production, it is the model-level (i.e. B1) that attracts our primary interest. We will therefore in the next section outline some characteristics of the COPICS concept.

Design features of COPICS

With reference to Figure 4.5 the COPICS pattern company and the COPICS model will be presented in this section. The emphasis will be on the underlying assumptions, here called design features, that constitute the basis for the COPICS program and which can be assumed to be crucial for the discussion of fit.

The COPICS pattern company

COPICS highlights three cornerstones of industrial activity, namely planning, implementation of the plans and control, which together define the essence of the COPICS concept. As described in the

COPICS manuals, the concept is developed as follows.

Planning implies the setting of objectives and defining policies and standards to meet the objectives. The essentials of planning constitute choosing from a range of alternatives.

Once the production plan for finished products (i.e. the master production schedule) has been established, its implementation results in purchase orders and shop orders. Production systems require rapid implementation of changes to the plan. For example, the delay of a component order can have far-reaching effects, which must be quickly communicated to production and purchasing departments so that priorities of related orders can be changed.

Effective implementation of the plan is helped by features such as:

- Purchase order procedures that reduce the amount of time between requisition and order release. At time of receipt, rush orders are automatically handled on an expedited basis.
- The ability to test the detailed effect of a single significant order on production capacity. This allows realistic confirmation of order delivery dates.
- Procedures to check that all required materials and tools are available before an order is released to the shop floor. This eliminates costly physical staging to check for shortages.
- Tool control and recall procedures that reduce shop order delays caused by tool shortages, ensure maximum tool usage before recall, and reduce scrap caused by worn tools or uncalibrated gauges.

In the COPICS pattern company, control requires standards both to measure performance and to provide incentives as well as to determine scrap allowances and work-centre queues. For example,

- Manpower-idle time can be minimized by modifying a factor that governs the planned queue of work at each machine centre.
- Part shortages can be controlled by altering factors that determine the amount of safety lead time allowed.
- Emergency downtime can be minimized by altering the factors used to calculate the preventive maintenance time interval.

Among the major design features we may thus summarize some which are assumed to be 'normal' features in an industrial environment:

- Production capacity regarded as a controllable parameter.

- Tool replacement and tool maintenance on an 'as-needed' basis.
- Reduction of time between requisition and purchase order release through improved purchase routines.
- Manpower-idle time reduction through better machine centre planning.
- Shortages control through the modification of safety lead times.
- Reduction of emergency downtime by modifying preventive maintenance schedules.
- The availability of standards.

The COPICS model

As the general design features of the previous section are narrowed down to specific application areas, they become objectives of the application. 'Purchasing' as a vital function of materials acquisition may here serve as an example. From the COPICS manual is quoted:

The objectives of purchasing are to obtain the required quantity of material by the date specified, at the lowest possible price consistent with the required quality level. To do this, purchasing must have time to evaluate an adequate number of suppliers for each item. For important items, current price and delivery quotations must be secured from several sources of supply. Care must be taken not to overload any one supplier, thereby jeopardizing on-time delivery.

Although price is important, so is the ability of the supplier to deliver within a short lead time. The requirement to place orders far in advance can result in higher inventory. Therefore, negotiating for shorter delivery lead time is also an important buyer function.

Purchase orders must be followed up to avoid late delivery and consequent upsets to the manufacturing plan. A supplier's capability to deliver on time must be evaluated, and delivery variations must be allowed for in planning.

Delivery of off-standard material can significantly disrupt manufacturing planning and execution. Therefore, the quality of each receipt must be measured against standards. Historical comparisons must be maintained to allow buyers to negotiate better performance from suppliers whose quality rating is drifting.

These objectives are interrelated. One supplier's low price may be offset by less consistent lead time, or by lower overall quality. Therefore, the factors must be weighed against each other. To do this, the purchasing department must maintain data on what items the supplier can furnish, his available capacity, and his historical price and delivery performance.

The techniques presented here allow more effective evaluation and selection of suppliers and reduce the clerical effort that consumes most of a buyer's time, thus freeing him for negotiation with suppliers.

Delay of component orders can have far-reaching effects, and fast information is regarded as necessary by the production and purchasing departments so they can change priorities of related orders. It is therefore within the objectives of COPICS to determine the effect of a component order delay and to dispatch messages that alter the priorities of work-orders in the other affected departments. Components not yet needed will therefore not be produced. It can also report which, if any, customer orders will be delayed.

A closer monitoring of activities would, according to the COPICS documents quoted, result in better performance, but an individual's span of attention and control must be regarded as limited. The system, however, can monitor a large number of widely dispersed activities and issue an 'alarm' when conditions require management intervention, such as when:

- An order fails to arrive on the delivery date specified.
- A bottleneck production facility goes down.
- An excessive amount of scrap is reported.
- A work centre is about to run out of work.

Alarms of these types can be generated as a result of normal manufacturing activity reporting. For proper action to be taken in response to such out-of-line situations, two 'control aids' are needed:

- A set of standards against which activity can be compared by the system to see if an abnormal condition exists.
- Action files, which are used by the system to direct the alarm message to the person who can correct the off-standard condition

These two features apply to all areas of COPICS.

The COPICS model, however, seldom considers to what extent individuals have the authority to react and take measures, to what extent the computer initiated directives are semantically coherent with the language used in another culture or to what extent implications of actions, which are often implicitly assumed by the system, are realized and understood by the employees. These and similar issues will be of significance in the following discussions.

Part II

From model to reality – Egypt

Um zu erkennen, ob das Bild wahr oder falsch ist, müssen wir es mit der Wirklichkeit vergleichen.
(In order to discover whether the picture is true or false we must compare it with reality.)

Wittgenstein: *Tractatus Logico-Philosophicus*

The overall purpose of this second part is to describe the production system in a Third World industry. The description will then serve as a frame of reference for our discussion in the third part, about how reality is interpreted into models, how the models are transformed into computer programs and, finally, to what extent these programs are applicable in different contexts.

We have already noted that the description of one particular organization is not a general description and the findings that can be observed from such a description are not generally applicable to other organizations. The wish to generalize results is further hampered as we realize that the terms 'developing countries' and 'developing industries' are shorthand concepts for different levels of economic, technical, social, cultural and political development. The great disparities between, and even within, developing countries make any attempts at generalization hazardous, sometimes also misleading. But with some caution we may identify a number of factors which apply to other developing countries as well and which also have impact on our model–reality discussion.

The company of interest for our empirical study is located in Egypt. Therefore in line with our earlier notion that it is only by knowing something abut the environment that we can fully understand the behaviour of an organization, the first chapter of this second part will be devoted to an overview of Egyptian industry.

Egypt as a setting for the empirical study is highly relevant. Egypt is a developing country with a large public enterprise sector that is entrusted by the political elite with the development plans of the country. The country is and has been for a long period faced with many social, economic and political problems, and it has been regarded as politically crucial to foster economic development programmes in order to ensure both economic growth and a more even income distribution. A political step in this direction was the introduction of a set of socialist laws in 1961. Despite the advent of the open door policy in 1974 that reopened the country to foreign business investments and facilitated the establishment of private enterprises, a large public sector with dominating control structures over industry is still a major feature in Egypt.

It is now interesting to study the applicability of computers in Egyptian industry for several reasons. The first is very pragmatic: industrialization is a major development strategy in Egypt, as in many other developing countries, but is confronted with a variety of obstacles and problems of an economic, organizational and technical nature. Attempts are continually being made to improve the situation and in the efforts to improve industrial effectivity, computers have been adopted for the support of specific key functions such as material requirements planning. It is therefore of interest to determine whether and to what extent the use of computers has contributed to improved production management in this developing industry.

The second reason is of a more general nature. Computer software such as application programs for production management can be regarded as a set of formalized routines, developed under assumptions of what constitutes, for example, a rational organizational behaviour or appropriate information practices. These assumptions, as we have seen, are based on a conception or model of reality and the computer software is based on a model where the formulation of production management problems and the solutions to those problems reflect a model of a production environment that we call a Western model. In Egypt, however, the structure of society, the economic and technical conditions and the many diversified and non-standardized ways of solving problems may require different models.

The inevitable question we pose is to what extent a computer system for the planning and control of production, designed and built around assumptions of rationality, in a Western sense, regarding production goals, control structures and information practices, is applicable in a developing industry, operating within a context that is largely dominated by non-Western traditions and values.

The question thus raised will be the departure point for the following. With this aim in mind, the perspectives, study levels and

aspects applied have been chosen with the sole objective of identifying phenomena and situations which influence the computerization process. It is therefore inevitable that other aspects and
perspectives have been omitted, neglected or only rudimentarily
considered. The interaction between the environment and the company
can be taken as an example. This interaction is naturally more
complex and multifaceted than it appears in the study.

The empirical material that is presented in the last three chapters
of this second part, is based on a study of EL NASR Automotive
Manufacturing Company (NASCO), a public company in the Egyptian vehicle industry (Lind 1988).

The focus of the study is the management of the material flow,
from material needs specification to input for final assembly. This
view will involve the essentials of material management problems
such as the fluctuation of procurement and fabrication lead times
and the existence of multi-sourced parts. These are difficulties that
frequently confront manufacturers in developing countries.

Focus on the material flow offers a perspective that relates the
company to its environment and, through this, implicitly involves
features that can be assumed to be characteristic for an organization
in a developing country. The relevance of this perspective has also
been emphasized by Pfeffer and Salancik, who claim that a good deal
of organizational behaviour is understood only by knowing something about the environment of the organization and the problems it
creates for obtaining resources. They state that:

> despite the importance of the environment for organizations,
> relatively little attention has been focused there. Rather than
> dealing with problems of acquiring resources, most writers have
> dealt with the problem of using resources.
>
> (Pfeffer and Salancik, 1978, p. 3)

Chapter five

Industry in Egypt

Egyptian industry can be characterized by two dominant factors. First, there is the large share of government-owned enterprises, contributing 75 per cent of all industrial output and including all medium- to large-scale industry. Second, the market for this public industry is a highly regulated market (beside the much less regulated intermediate market) that is partly deprived of its self-adjusting mechanisms through state-directed price levels. Industrial policy and strategies, on the national level, are visible in the individual public enterprises, shaped to be part of the on-going process of political change and development.

Examples of such policies and strategies are the attempts to reduce the foreign exchange gap through fostering input substitution in commodity import (El Dabaa, 1983) and the prescriptions applied to public enterprises in order to increase employment. The overstaffing that follows from this policy is socially justified but financially and organizationally burdensome for public enterprises.

The static bureaucratic organization in government bodies influences work within and between the public enterprises. As witnessed and testified by several writers (Issawi, 1982, Ayubi, 1982), slow decision procedures and lack of local authority in enterprises (Badran and Hinings, 1981) impinge upon local initiatives, complicate utilization of allocated resources and slow adaptation to change.

As a public enterprise, NASCO is subjected to state control, both directly through steering committees and indirectly through currency allocation and price setting. A more detailed analysis of the company is therefore complemented by the following survey of the industrial environment.

The industrial background

As pointed out by several writers (e.g. Abdel-Khalek, 1982, Amin, 1983, Ayubi, 1982), the present industrial (and economic) situation of Egypt has many parallels with Muhammad Ali's Egypt of the early nineteenth century. Issawi observes that:

> An early and vigorous attempt to establish modern industry was made by Muhammad Ali in Egypt in the 1820s. By 1838, investment in industrial establishments amounted to about L12 million, and employment to some 30,000 persons, an impressive figure in a total population of about 4 million. The industries covered a wide range, including cotton, woolen, silk and linen textiles; dyeing; foundries; sugar refining; glassware; tanning; paper; chemicals; arms and ammunition; shipyards. But Muhammad Ali's factories, largely designed to meet military needs, were kept going only by his enormous energy and constant supervision, and suffered from great inefficiencies, including lack of fuel and metallic raw materials and the total absence of skilled labor, which meant that not only foreign engineers and supervisors but also foreign workmen had to be brought in. Production was generally well below capacity and spoilage was considerable. Moreover, the factories survived only thanks to the high protection provided by the monopolies.
>
> (Issawi, 1982, p. 154)

A common feature that can be observed in the situation now and then is the strong attempt by the government to exert control over industrial activities. In a survey of contemporary literature in Islamic economics, Siddiqi (1976, p. 208) refers to several writers who stress the role of the state and its direct involvement in financial and other matters related to industrial development, where the role of the state as a legislator is seen primarily as a complement to Shariah, the Islamic law. However, as individual industries need to formulate and implement their own control strategies, there is a considerable risk of a functional conflict between the state on the one hand and the individual enterprises on the other hand, since goals and time-frames tend to be partly incompatible.

State involvement

The formal tools of the government for exerting control of the development process in today's Egypt are the five-year plans that started with the 1960–1 to 1964–5 plan. Targets like boosting industrial development, a common goal in most of the plans, result in a variety of different measures such as adjusting import/export tariffs and setting custom duties. A further control tool is the currency exchange rate, officially set at 0.7 EGP for one US$, but which now has competition from a higher informal (or free) exchange rate, 2.1 EGP for one US$; as a matter of fact, not less than four different exchange rates are in use, each applied to particular types of transactions, commodities or enterprises.[1] This mixed currency situation severely restricts the government's prospect of controlling the country's finances.

From an organizational point of view, the central government is involved through the President, who has authority to give public status to a company and to determine the percentage of net profit to be reserved for the purchase of government stocks or for deposit in a special account at the Central Bank. In addition, the Prime Minister has power to appoint the managing board's chairman and half of its members and to add others. The responsible minister has extensive powers over the enterprise, such as assigning targets, evaluating performance and suspending the chairman and board. Central government, then, has the job of policy guidance within the terms of the national plan as well as the power of appointment and suspension of board members. Moreover, bodies such as the Central Agency for Audit, for Organization and Management, for Mobilization and Statistics, the Administrative Investigation Boards, and the Ministries of Finance and Planning all have the right to inspect and to demand specific information. This administrative control is also accompanied by juridical and political control over the enterprise (Badran and Hinings, 1981).

The state involvement is in itself a dominant and important background factor in the industry and constitutes a solid structure, strongly tied to Egyptian society and through the social and cultural rules also to the Islamic community. Within the structure, industrial policy and strategy decisions and issues will constitute a background for NASCO. This background can be assumed to be both complex and many faceted. In the following, we will attempt to discern one background pattern of relevance for this study, one chosen perspective out of many.

A background pattern

The industrial situation in Egypt, dominated by the public sector that accounts for about three-quarters of industrial output and even more in annual industrial investment, is tightly tied to the general economic conditions of the country, which can be characterized as a deficit economy, regardless of which measures are used.[2] The trade balance, for example, is given in Table 5.1.

Table 5.1 Foreign trade 1981–5 in millions of US$

	1981	1982	1983	1984	1985
Export	3999	4018	3693	3864	3836
Import	7918	7733	7515	9250	8338
Trade balance	–3919	–3715	–3822	–5386	–4502

Source: ABECOR reports April 1985 and June 1987

The considerable deficit in the government budget is due to higher growth in total expenditure than in revenue.[3] This is partly explained by a slowdown in some of the major sources of national income, e.g. remittance from Egyptians working abroad and the oil export. Of importance also is the increasing cost of the food subsidies that were introduced to reduce the impact of fluctuations in world prices. This dependency is in turn largely caused by a shrinking agriculture sector which has led to an increased import of foodstuffs, now amounting to over 50 per cent.

The government deficit (budget and trade) is now largely financed by foreign aid and loans and through the banking system, with the consequence that funds and savings which have been ear-marked for productive investments are used for current expenditures such as food subsidies and to maintain other artificially low price levels. It has thus been estimated that a fall of, for instance, 10 per cent in Egyptian foreign exchange availability results in a fall of 4 per cent in industrial output and a 6 per cent drop in industrial investment. And, if the prices of imported food go up by 10 per cent, industrial output falls by 1–2 per cent and investments by 2 per cent (*Economist*, April 1984).

In parallel to this, the foreign debt that, according to recent IMF figures, amounted to US$32 billion in 1984, is projected to increase to US$50 billion in 1990.[4] The growing debt has recently downgraded Egypt to a high-risk category in various international financing organizations,[5] although short-term credits are largely being repaid on time. These credits, however, are more and more being used for the purchase of foodstuffs, thus leaving less over for the acquisition of other vital commodities such as raw material and machinery.

A factor that has crucial impact on the allocation of financial support for long- and short-term investments is the bureaucracy, as testified by several writers. Ayubi (1982, p. 369) thus refers to bureaucracy as a hampering factor for the utilization of available aids and loans, and Issawi states that:

The huge expansion of the public sector has brought with it many problems. In all countries [in the Middle East] four or five year plans have been drawn up and over the last two decades the planning process has improved somewhat. But as in other parts of the world, the government machinery is not equipped to handle the immense sums involved – which in all countries run into many billion dollars and in several into the tens of billions. Again as in other places, with the great expansion of the bureaucracies the quality has deteriorated, the energy provided by the few highly qualified, dedicated men at the top being dissipated by their subordinates. With the growth of government expenditure and intervention, corruption has flourished. And the concentration of economic, social, and cultural activity in the government has further undermined the foundations of democratic self-government, which at their best were weak.

(Issawi, 1982, p. 184).

The priority process gives rise to shortages of various kinds: raw material and machinery as already mentioned, but also spare parts for imported machinery and transport equipment. On the other hand, and as a contradiction, there is an obvious overcapacity in industry, reported to be as high as 30–40 per cent (Issawi, 1982, p. 168). In this perspective it is perhaps more relevant to use the term capacity imbalance than capacity shortage.

Industry, public as well as private, using machinery of foreign origin (and practically all machinery is imported) and imported material, is in need of foreign currency for import. Public companies, being exposed to increasing waiting times for foreign exchange and, at the same time, to the government's industrial improvement policy, are in constant need of raw material, machinery and spare parts. Expensive foreign currency therefore has to be bought on the free exchange market, resulting in an accumulated government debt to public industries that by 1982 had reached 500 million Egyptian pounds.[6]

Shortage of foreign currency therefore contributes to the imbalance of capacity and increases vulnerability to production interruptions and breakdowns. This situation is worsening, and there is evidence of a growing use of imported inputs in industry.[7] The present five-year plan is therefore encouraging a reduction in import dependency.

The focus on import substitution in industry is frequently referred to as a major government development goal, and different measures are taken to promote this goal. But import substitution has hitherto primarily affected consumer goods industries and there is a risk that import substitution will encounter difficulties once the easy replacement of the limited range of consumer good inputs has been achieved.

A UNIDO report (IS.472, 1984) on the automotive industry, states that the emphasis on local content of parts fabricated locally instead of being imported, i.e. import substitution, is not simply a question of quantity. As new components and material, new production processes and new types of machinery appear in the developed countries, developing industries will find it easy to argue for a reduction in local content and the import substitution strategy 'will continue to be a moving target which is never hit' (p. 131). It must therefore be assumed that an import substitution strategy that aims further than just easy replacements will generate a need for new facilities such as more advanced technology and new skills or, as pointed out by Handoussa (1986):

> Import substitution would also involve increased rather than diminished reliance on the import of the necessary intermediates to operate those industries.

Stewart states in this context that:

> 89 per cent of South imports of machinery and transport equipment come from North countries, 6 per cent from centrally planned economies and only 5 per cent from other South countries.
>
> (Stewart, 1978, p 122)[8]

This is further underlined by UNIDO (ID/304, 1983, p, 290) in stating that:

> the users of machine tools in those [developing] countries are heavily depending on suppliers in developed countries.

The import of technology and other facilities required to achieve import substitution therefore leads to a further outflow of capital which in turn further contributes to the deficit in hard currency. The attempt to picture the chosen background pattern of the Egyptian economy is summarized in Figure 5.1. In order to break this 'vicious circle' a major strategic goal is to reach a domestic manufacturing output level at which the costs for the import of capital goods, which is necessary to reach this manufacturing output level, will be compensated. Increased import substitution together with increased export of industrial products would thus result in a more balanced national budget.

In connection with this various writers do, however, outline a rather gloomy picture of recent development in Egypt. Daghistani

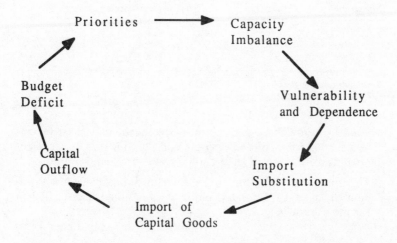

Figure 5.1 Causes and effects in the Egyptian economy: a background pattern

(1985) thus concludes in a comparative study of Egypt and a number of other developing countries that:

> The expansion of the industrial sector in Egypt did not have as much impact on the import substitution or exports of manufactured goods as it had in other sample countries.

Import substitution and industrial export tend to partly conflict in the sense that scarce resources such as capital and skilled labour, which are vital for maintaining a competitive strength in the export sector, are now being re-distributed to the import substitution industry. Export is further penalized by the over-valued exchange rate and by the fact that outright prohibition of certain competitive imports has resulted in low efficiency and quality among local suppliers (UNIDO, 1983, ID/304).

The situation can be illustrated by Tables 5.2, 5.3 and 5.4. The first table shows the annual growth in capital goods import compared to total import for the period 1966–79. The table shows that the share of capital goods as imports has grown substantially in the period since 1974 that marks the beginning of 'infatha' or the open door policy, a policy adopted to foster the industrialization of Egypt. As

Table 5.2 Average annual growth in percentage of total import and capital goods import

Time period	Total import	Capital goods import
1966–1970	–6.3	–5.2
1970–1974	36.3	19.6
1974–1979	19.5	38.3

Source: Data compiled from CAPMAS.

testified by many observers, the most significant impact so far has been a substantial increase in private consumption and in consumer goods industries rather than in capital goods production.

The open door policy for foreign investment announced by President Sadat in 1974 created more 'fat cats' than it did efficient Egyptian business.

(*Business Week*, September 1985)

As discussed above for the vicious circle to be broken, the substantial growth in capital goods import ought to result not only in comparative growth in industrial output but, in particular, in manufacturing output; manufacturing being the most dynamic and hence most significant part of the industrial sector (Bairoch, 1975, World Development Report, 1982).

The following two tables do not encourage such an interpretation. First, we see from Table 5.3 that the growth in manufacturing output, measured in value added, for the three time periods of Table 5.2, was less than the growth in total industry output for the same period.

From Tables 5.2 and 5.3 we may now define an 'elasticity factor' between capital goods import and manufacturing output. This elas-

Table 5.3 Average annual growth rate in total industry and manufacturing output in added value

Time Period	Total industry	Manufacturing
1966–1970	5.5	8.7
1970–1974	11.7	7.5
1974–1979	23.2	14.6

Source: Data compiled from the *UN National Accounts Statistics*, 1981.

ticity factor is simply taken as manufacturing output growth divided by capital goods import growth, i.e. the elasticity factor q is defined as:

$$q = \frac{\text{growth in manufacturing output}}{\text{growth in capital goods import}}$$

Table 5.4 Elasticity between manufacturing output and capital goods import (from Tables 5.2 and 5.3)

Time period	Elasticity factor q
1966–70	–1.67
1970–74	0.38
1974–79	0.38

Except for the first period (1966–70) that gave a negative elasticity (i.e. manufacturing output grew in spite of a reduction of capital goods import), there is a fairly stable elasticity for the period 1970–9. For this period, Table 5.4 shows that a 1 per cent increase in capital goods import resulted in only around 0.4 per cent growth in manufacturing output or, in other words, growth in manufacturing output in the period 1970–80 could not compensate for the much higher growth in import of capital goods.

The import substitution policy of Egypt has a direct impact on public industry, and hence also on NASCO, as we will notice in the following. The benefits of this policy in a wider development context are, however, much debated, and there is great variation in the opinions held by different writers.[9]

The industry structure

Practically all medium- to large-scale industrial enterprises in Egypt belong to the public sector. El Dabaa (1983) reports that there are approximately 200 public industries, with a total employment of about 600,000. The private sector has more than 150,000 small establishments, of which only about 3 per cent are factory type operations employing more than 50 people. In general, the private sector has been limited primarily to textile manufacturing, food and beverage processing and the processing of wood products, light metal fabrication and leather products.

A listing of public enterprises in size-order is somewhat uncertain as data are unreliable. Compiling data from different sources, however, gives a fairly good concordance for the ten major enterprises (see Table 5.5). Again, a word of warning as to data accuracy. In another survey (*South*, May 1986) the ranking of companies is partly changed.

Table 5.5 The ten biggest Egyptian public enterprises

Company	Turnover (mUS$)	Employees
Suez Canal Company	957	n.a.
El NASR import/export	735	1500
Egypt Petroleum Co.	680	35400
Egypt Air	499	n.a.
Eastern Tobacco	442	9400
NASCO	300	12000
Egypt Iron & Steel	258	24000
Misr Spinning & Weaving El Mehalla el Kobra	214	34000
ACOTAR Cotton and Trade	72	1000
Biscomisr	60	4000

Source: South, November 1985; data compiled by the author.

Computerization in Egypt

In 1962, the first computers were introduced into Egypt and by 1979 the number of computers had reached 120 in government and the public sector, and 185 if the private sector is also included. The first computer to be installed at a university was in 1964 (at the University of Alexandria), followed by Cairo University and Ain Shams University in 1968–9.

With ICL established in Egypt already in 1934, albeit under a different name, only a few more computer vendors were entering the Egyptian market in the early period of 1963–74, namely, IBM and NCR. With the beginning of the open door policy in 1974–5, a number of computer vendors were attracted by the opening Egyptian market and the number of vendors increased rapidly. One official source reports that more than thirty firms were represented on the Egyptian market in 1981, another reports twenty-two firms. It should, however, be noted that NCR, ICL and IBM, together still represent nearly 60 per cent of all installed computers. Of the rest, more than ten vendors have sold only one or two systems each which is often too few for the critical volume of establishing a service organization and other kinds of support to the users.

A figure indicating a general level of performance and effectivity of computer usage in Egypt is not available although low utilization and difficulties to financially motivate many installations is obvious. According to one source, 70 per cent of all installed computers can be regarded as not optimally utilized. Another source claims that it is not unusual to have an effective utilization of 5–10 per cent and it has even happened that computers are unused for 2 and 3 years.

A general conception is that about 10 per cent of all installed computers can be characterized as well utilized, efficient and achieving their objectives, whereas 90 per cent are regarded as inefficient in one way or another.

Major application areas which have been developed are accounting and general ledger, payroll, and basic costing and inventory accounting (stock control). Data processing is primarily batch-oriented (no terminals) and only a few organizations – mainly in banking and tourism (e.g. Egypt Air) – have been able to move towards real-time database systems.

Industry (mainly public) like textiles, automobiles, iron and steel and chemical (including petrol) have adopted computers but primarily for administrative applications as above. In the iron and steel sector, however, computers are in use for on-line maintenance management as well as for process control and in the automotive sector computer-based solutions for materials management are being tested (NASCO).

The relatively low ambition as regards industrial applications is because the computer suppliers (and software houses) do not take any particular initiatives to promote their industry products as the market potential is considered very limited and this type of application in general requires substantial support, before as well as after its installation.

Chapter six

NASCO – a company presentation

After about one hour's driving from Cairo, on a constantly jammed highway along the Nile bank, one reaches the town of Helwan, 25 kilometres south of Cairo.

Overlooking the Nile to the west, and with the ruins of Memphis and the Saqqara pyramids in the background, Helwan is another one of the many meeting places between ancient Egypt and today's industrializing society: the prestigious Helwan iron and steel works, the fertilizer plant and the numerous cement plants. Here, on the borderland between the fertile strip of river bank and the desert in the east, El Nasr Automotive Company (NASCO) was established in 1960, in accordance with governmental decree number 913.

Today the company occupies an area of nearly one million square metres, and with a work force of 12,800 in 1984 produced around 27,000 cars and commercial vehicles.

NASCO is a significant company in Egypt, with respect to its size, its products and its technical level. In turnover, it ranks number six in the country, number two in production and number one in manufacturing.

The company

A historical framework

In the period from 1952, the advent of the republic, until 1956, there were no substantial institutional reforms in the country that affected Egyptian industry. In 1956 the first separate Ministry of Industry was established, and in 1957 the Economic Organization was set up to function as a holding company for mixed (state–private) companies and Egyptianized foreign property. By 1958, this organization held control capital in fifty-two companies. In 1961, during the nationalization era, and as a step in the dramatically expanding public sector

(Abdel-Khalek, 1982), the MISR and the NASR Organizations were established, alongside the Economic Organization. Shortly afterwards, the three organizations were split into forty organizations on the basis of types of activities. Until the middle of the 1970s, the beginning of the open door ('infatha') policy, the organizational structure of the Egyptian industrial sector was based on control by the appropriate governmental organization between the ministry and the individual company. Abdel-Khalek states that:

> The public enterprises were considered the 'private' property of the state, and they were manned by individuals who more often than not had no experience in the field, while the organization of the public sector failed to provide enough checks and control against illicit practices by the management of the public enterprises.
>
> (Abdel-Khalek, 1982, p. 262)

In 1975, public organizations were therefore abolished by a law that concentrated power in the hands of the management boards of the individual public companies. This power gave the management board responsibility for following up the execution of projects and meeting deadlines, achieving financial targets, developing resources, recruiting and training manpower and implementing quality and quantity control with a view to making the best use of resources (Badran and Hinings, 1981).

The government's need to plan and control the public industry has recently resulted in a new umbrella organization, the General Organization for the Public Sector with sub-divisions for individual branches (engineering industry, chemical industry, etc.). The new structure adheres to a traditional pattern of governmental interferences, justified, from the government's point of view, by inadequate management capacity and by the need to balance, on a national level, demands with availability of scarce resources.

But the execution, at a company level, of plans that have been established at a government level, leads to misfit between *is* and *ought to*, between what is possible and what is wanted. This also has, as we shall see, an impact on the use of computers for planning at a company level.

It is within this framework that NASCO came into operation in 1960 and, since then, has developed.

Basic company figures

A very concentrated picture of NASCO for the fiscal years 1982–3 and 1983–4 is given in Table 6.1.

Table 6.1 NASCO's performance in 1982–3 and 1983–4

	1982–3	*1983–4*
Total revenue ('000)	260,232 EGP	320,000 EGP
Wages ('000)	30,730 EGP	36,696 EGP
Employment	12,400	12,800
Production of		
trucks	2,871 units	3,198 units
buses	701 units	693 units
tractors	3,294 units	4,493 units
cars	20,896 units	19,033 units

Source: Automotive Industry in Egypt, 1985. (1 EGP is approximately 1.2 US$ at current rate)

The organization

NASCO is organized into six sectors, each sector having a number of divisions (see Figure 6.1). The strong ties to government are indicated with a dotted line to the Ministry of Industry. For the following discussion about organizational control and applicability of computers it should be stressed, again, that in developing countries political objectives cannot be ignored because of its strong influence on public organizations as we have seen above.

The mission of NASCO

In addition to producing vehicles for the Egyptian market and, if possible, for export, an essential mission is to promote and enhance domestic production by encouraging local enterprises. There are two main reasons for this: to build an indigenous skill in industrial production and to reduce the outflow of capital for the import of material and parts.

NASCO has a vital role in the establishment of local industries, both through its feeder industry department and in other joint projects, and there is a continuous effort to exploit any available potential to increase domestic content in the final products. This mission has a direct impact on the computerization process of the company, since the goal to increase local content also applies to its own production, where parts of production increase relative to assembly. This creates a demand for improved co-ordination of production and material supply and, hence, for efficient production planning and control methods.

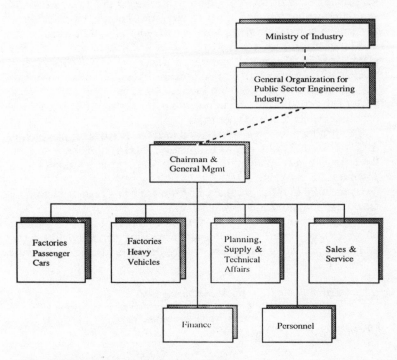

Figure 6.1 The NASCO organization: six sectors

Products and market

The products

In 1959 the first agreement was signed for implementation of a project to produce trucks and buses under foreign licence. The initial programme of civilian medium trucks with a payload of 6 and 8 tons and buses for 48 passengers was soon followed by production of heavier trucks and buses for 110 passengers. In 1983, 2,871 trucks and 701 buses were produced, with a planned increase for 1984 of 25 per cent and 7 per cent respectively. (Actual increase was 11 per cent for trucks and a decrease of 1 per cent for buses).

The local content of components in the trucks and buses has gradually been increased to reach a value of 65–75 per cent, less however in the volume of components. Locally produced parts are roughly split 50-50 between in-house and local production by feeder industries. The remaining 25–35 per cent represents parts which can not be economically or for other reasons produced in Egypt and which are therefore imported. Such parts are, for example, electrical components, fuel injection pumps and some types of ball bearings.

Manufacturing of trailers started in 1961 under licence from Blumhard Co. (Germany), and the product programme comprises of models with a payload of 6 to 26 tons. In 1983 the trailer production was 364 units with an expected increase of 10 per cent the following year. (The actual result was a decrease by 47 per cent.) Value of local components content is 70 per cent.

The production of tractors started in 1961. The tractor production for 1982–3 was 3,294 units, with a planned increase of 20 per cent for the following year. (Actual increase was 36 per cent.) Value of local content is 25 per cent.

In addition to these products, engines are being produced under a mixed manufacture/assembly scheme.

A summary of NASCO's production in terms of average number of vehicles produced per year over three time periods, and for the four main product groups, cars, trucks, buses and tractors, is shown in Table 6.2.

Table 6.2 Average NASCO vehicle production 1960–83

| Vehicles | Average production per year in | | | |
	1960–9	*1970–9*	*1980–2*	*1983*
Cars	1,240	8,575	15,845	23,555
Trucks	730	1,316	2,244	2,871
Buses	224	370	600	701
Tractors	744	1,848	2,116	3,294

The present (1985) programme for the heavy vehicles division has capacity for the following volumes:

Medium trucks:	20–25	units/day
Heavy trucks:	7	units/day
Buses:	3–4	units/day
Tractors:	20–25	units/day

The daily capacity for trucks in 1985 was thus about three times the average production of trucks in 1983, a capacity that was not utilized.

The market

The Egyptian market absorbed in 1982 around 70,000 passenger cars and commercial vehicles. Of these, a little more than 40,000 were imported and the rest locally manufactured and/or assembled. The estimated market for 1990 is an annual demand of around 100,000 cars, 5,000 buses (medium/heavy) and 16,000 trucks (medium/heavy). To this can be added a substantial further market potential in the North Africa and Middle East region as well as in Africa south of Egypt. Egypt as a major centre for North African markets has been considered by Japanese as well as European and US automotive manufacturers (Issawi, 1982, UNIDO, IS.472, 1984).

Studies of the world distribution of motor vehicles show the importance of per capita income as an explanatory variable in the patterns of ownership (Bloomfield, 1978). This is also true for the Middle East countries and explains why motor vehicle density in the region and the distribution of commercial vehicles (trucks and buses) and passenger cars reveals quite considerable variations between the countries.

From Table 6.3 we can see that the Egyptian market, from a statistical point of view, has absorbed less vehicles in total than the other countries, and that the regional market for commercial vehicles seems to be under-exploited. For NASCO's heavy vehicle production the outlook therefore ought to be promising.

Table 6.3 Density and distribution of cars and commercial vehicles (1987)

	Persons per com .v.	*Persons per car*	*Cars %*	*Com. v. %*
Egypt	211	125	63	37
Saudi Arabia	10	11	47	53
Kuwait	9	3	74	26
Jordan	43	20	68	32
Syria	61	120	34	66
Regional average	53	31	63	37

Source: World Automotive Market 1989. 'Regional average' indicates calculated average figures for the whole Middle East region, i.e. for more countries than those shown in the table.

NASCO – a public company

As already noted, Egyptian industry is characterized by state regulation of the market and a large share of public enterprises. Government establishes rules, temporary or more long-lasting, which govern relations between a public company and the market or between public companies; rules that are justified from the country's perspective but not necessarily from that of the company. An example of this is transferring the purchase of components from a foreign to a local supplier. This may be regarded as important from the government's point of view (import substitution) whereas for NASCO the result may be quality problems or delivery delays.

In their study of Egyptian public enterprises from a management point of view, Badran and Hinings (1981) found that the locus of authority is largely outside the organization's own structure and that Egyptian public enterprises therefore exhibit lack of autonomy as well as a natural centralization of authority within the company. At NASCO this can be illustrated as follows. The objective of the company, according to high level managers in production, is to increase production of vehicles in order to reduce the need for imports, something which is also well in line with the official strategy of the country. The production of buses was taken as an example.[1] From the company management's point of view it was therefore regarded as a contradictory step by the government when negotiations with Daimler-Benz (West Germany) led to an agreement on an annual production of 600 buses, a number that corresponds to the total annual expected demand for heavy buses in Egypt in 1990.[2] It can be added that the NASCO production programme includes heavy buses.

Another illustrative example is the negotiation between the Egyptian government and General Motors (USA) regarding a joint venture for passenger car production. A GM newsletter of June, 1986 reads:

> General Motors has received permission from the Arab republic of Egypt to form a new joint venture in Cairo, Egypt, it was announced today by [name], Vice President in charge of Asian and African operations.
>
> The new company, to be named General Misr Car Company S.A.E. (GMCC), will participate in the Egyptian passenger car industry and will complement the two-year old General Motors Egypt S.A.E. which manufactures and distributes commercial vehicles.
>
> Equity in GMCC will be shared between General Motors (30 per cent) and Egyptian investors (70 per cent), including El Nasr Automotive Manufacturing Company (NASCO) with a 30 per cent share. Other shareholders include the Misr Iran Development Bank and the Export Development Bank of Egypt.
>
> According to [name], GMCC will import components for [type of cars] and distribute, sell and service these models in Egypt. GMCC will contract with NASCO and the Arab American Vehicle Company (AAV) to use their existing facilities for manufacturing the cars. Production is expected to begin in mid-1987.

[name] further noted that, in addition to managing GMCC, GM personnel would be responsible for product quality, material control, and production schedules, and would provide technical assistance to upgrade and re-equip the NASCO plant.[3]

As can be noted, the GM newsletter spells out far-reaching implications for NASCO in new business commitments and in a new product programme with new production procedures in vital functions such as quality control, material control and production scheduling. The new joint venture would also place requirements on NASCO management for adaptation to a new corporate culture and values.

In April 1987, almost a year after the GM newsletter, there was still no formal agreement from NASCO on participation in the joint venture. Indeed, NASCO's production management expressed great doubt regarding the benefits for NASCO of being a counterpart in the project. Again, the decision-making authority is located outside the organization, something that has a significant impact on the formulation of business goals.

These and similar examples illustrate the lack of consistency, between government and company levels, regarding the long-term objectives of the company. While change is desirable for the state, stability is necessary for NASCO. These two modes can be compatible as long as change takes place over a long enough time-frame. However, rapid change in the Egyptian economy is called for from several directions in order to improve the present economic situation. Change is therefore not compatible with stability on the company level and over a short time-frame this incompatibility leads to lack of co-ordination in daily activities and difficulties in establishing procedures and routines which are long-lasting enough to enable standards and references to develop. As pointed out by Youssef (1979):

> Prediction of future events depends to a large extent on describing present and past events. Therefore, prediction becomes difficult if the future keeps shifting and is a radical departure from the present.

As we will see in the following, this difficulty in predicting future events is a crucial factor in the planning of NASCO's production.

Chapter seven

Vehicle production at NASCO

The production of vehicles

The making of any type of motor vehicle involves in general three basic stages: design, manufacturing of parts and the assembly of parts, first into sub-sections and then into the complete vehicle.

A characteristic of production under licence, as with NASCO, is that design is very limited and in practice restricted to minor engineering changes imposed by local requirements. It is therefore correct to state that the two major production phases at NASCO are parts manufacturing and assembly.

At NASCO, stamping of sheet steel into body shells (see Figure 7.1) is the only basic process done by the company. Forging and casting of, for example, engine blocks and gear-box housings is done outside, partly at El NASR Company for Forging, another public industry in Egypt. For basic parts production, as well as for sub-assembly steps, there are thus different sources of supply: from foreign sub-contractors, from other Egyptian companies and from in-house production.

In the following we will restrict the study of NASCO production to the flow of material for final assembly. This view will, however, include not only material but also in-house manufacturing of parts and sub-assembly of components.

Production control

At NASCO the term production control is used for both manufacturing and assembly as well as for materials supply, in accordance with general practice. Practice is, however, not without ambiguity, as both meaning and content tend to vary among companies. Voris (1966), as an example, questioned whether material control should be a production control function or not.

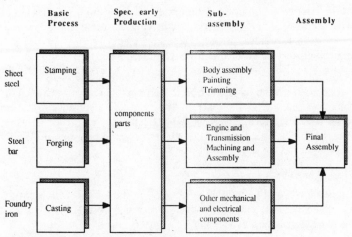

Figure 7.1 The motor vehicle production process
Source: Bloomfield,1978

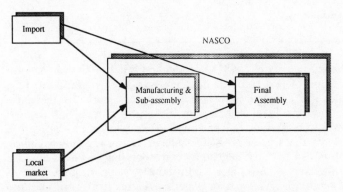

Figure 7.2 Multi-sourced supply of material for final assembly

In the following we will be using the term 'materials management'
to emphasize that manufacturing and sub-assembly together with external procurement constitute the three sources of material supply
for final assembly (see Figure 7.2). It is also the task of materials
management to ensure the supply of materials for in-house manufacturing.

A suitable definition of materials management is:

> Materials management ... comprises all the functions associated with the flow of material, from identifying the need to place orders on outside suppliers, through buying, receipt of goods, ordering of manufactured items and control of manufacture, to dispatch.
>
> (Corke, 1977, p. 245)

It is therefore not an exaggeration if we propose that materials management is the most vital part of the NASCO production system. That is also the reason why we will devote the whole of the next chapter to further analyse NASCO's materials management system.

Cost of production

Turnover for NASCO in 1982–3 was 260 million EGP which rose to 320 EGP in 1983–4. This turnover covered the production costs (and some losses) grouped as follows:

Table 7. 1 Distribution of production costs

material costs	60–80%
wages	10–15%
depreciation/interest	10%
overheads	the rest

The financial aspect of NASCO production is not on the whole within the scope of this study. It has, however, been pointed out (e.g. Hermele, 1982) that despite favourable salary levels for the automobile industry in developing countries, a rational production requires a substantially higher volume than what is produced in a company of the size of NASCO. Due to low production volumes, production can not be balanced by corresponding revenues in sales. Again, however, NASCO as a public industry is a part of the government's long-term development planning. Terms like profit and loss are therefore not necessarily the most relevant efficiency parameters.

Cost of material

The average cost of material for NASCO was 200 million EGP in 1984, and is thus by far the dominant cost factor. Even minor variations in material price and exchange rates therefore have a significant impact on the total costs, not least as a large part of material is imported and currency for materials import is sometimes obtainable only at a high price.

Imported materials, as well as imported production equipment and spare parts, are subjected to government regulations on import duties and exchange rates, and prices are subject to industrial policies. Local prices for parts and material can therefore be kept (artificially) higher than corresponding international prices. El Dabaa (1983) in his study on cost/benefit analysis of a caustic soda plant in Egypt thus states that:

> every item of indigenous production, no matter how costly, is protected from competitors through industrial licensing, import quotas and protective tariffs. The result of this policy is to divorce market determined investment decisions from any guidelines that international opportunity costs might otherwise produce.

El Dabaa concludes in his study that Egyptian local prices for construction material are 20 per cent above the international price level. Similar observations are not unusual from developing countries. Brodén (1983), for example, reports from Tanzania that locally produced batteries for automobiles were three times as expensive as imported ones.[1]

Cost of production equipment

It is frequently reported in Egypt that in spite of the growth in production volumes, the financial situation of public industry remains weak.[2] The major part of funds that are made available from the government budget for use in the public industry is consumed for the purchase of material for production. For additional financial means, required for investment in production facilities such as new machinery, spare parts, etc., companies like NASCO are compelled to approach the open money market where money can be borrowed, albeit at a higher rate. The following example shows the consequences.

Using government-provided money for the acquisition of spare parts with a price of US$ 100 requires 70 EGP in the official exchange system. With the unofficial rate from the open market the same spare parts require 215 EGP, as the rate is 2.15 EGP instead of 0.7 EGP to the US$. Spare parts, and in particular new machinery, are therefore very costly investments for NASCO, something that is reflected in the age distribution of equipment in use (see Chapter 8).

Salary costs

Average monthly salary for labour at NASCO is 100 EGP. To this is added an incentive that is calculated as follows: an incentive factor,

77

updated every second or third year, is multiplied by the gross monthly turnover in EGP. The resulting factor, called incentive days, is then multiplied by the daily salary and the result is added to the monthly (fixed) salary. The salary is in general substantially increased by these incentives.

The incentive system is more a government procedure to exercise welfare than a tool for NASCO to improve productivity, as it is tied only to revenue and not to production costs (or profit). The effect can be illustrated through the following example.

Assuming a production cost of 10,000 EGP for a particular product, the share of imported material may be 4,000 EGP. With a sales price of 11,000 EGP the profit is 1,000 EGP. This, however, is based on the official exchange rate of 0.7 EGP to the dollar. As NASCO is often forced to use the open money market (see above), foreign currency may be available only at the rate of 2.15:1. The cost of imported material is thus increased by 8,200 EGP and profit turns to a loss of 7,200 EGP. With unchanged revenue and with the same amount of production work performed it is, however, difficult to find convincing arguments, from the employees' point of view, for reduced incentives.

In the period 1961–2 until 1983–4, salary costs increased by a factor of 100. In the same period the number of workers increased by a factor of 5. In other words, the relative salary level for workers at NASCO increased by a factor of 20 in the period, a factor that would still be significant even if the effect of inflation was taken into consideration.

Production efficiency

I do not intend to go into any deeper discussions on production efficiency and productivity. However, as the concepts of productivity and productivity increase are often used to justify an investment in computers, it may nevertheless be relevant to make some comments.

A very simple measure used to describe growth in the automobile industry, used by, for instance, Bloomfield (1978), is the ratio between number of vehicles produced and total employment. In a comparison with the historical growth in the US automobile industry, the ratio for NASCO in 1982–3 (see Table 6.1) was reached around 1910–12 in the USA. Due to overstaffing of Egyptian public industry the comparison is, however, not quite adequate.

The productivity measure commonly used at NASCO is the value of total production divided by total wages per period. The ratio for

Table 7.2 Ratio of total production value to wages

Time period	production value/wages
1980–8	9 : 8
1981–2	7 : 2
1982–3	7 : 8
1983–4	8 : 7

the first half of the 1980s is shown in Table 7.2. An interpretation of the table could be that 1 EGP spent on wages yielded a return of 8.7 EGP for the year 1983–4. This interpretation is, however, misleading as a productivity measure, since value of production is determined primarily by cost of material (see Table 7.1).

A more common measure for productivity is therefore based on added value (output) and capital and labour (input) (Sumanth, 1984, Lind, 1985b). Figures on added value for NASCO production can, however, be estimated only with a high degree of uncertainty, making their usability very limited. Capital in Islamic economics is, as pointed out by Kazarian (1986) and Uzari (1976), not regarded as a separate production factor but rather as part of production at large. Uzari states that:

> A postulate which is essential in the analytical framework of Islamic economics is that capital as a separate factor of production does not exist but is rather a part of another factor of production, namely enterprise.

> (Uzari, 1976, p. 38)

It is therefore difficult to have a discussion in terms of tied-up capital and, hence, also of capital productivity as well as return on investment, as cost of capital is not a well defined concept in the NASCO context.

Labour productivity is also difficult to assess. Abundance of labour has led to overstaffing, with a direct negative impact on productivity measures. Also in spite of the abundance of labour, productivity is further impaired by lack of skilled workers.

Computerization of production control at NASCO

NASCO is among the pioneering companies using computers in Egyptian industry. A number of applications have been implemented for administration (e.g. payroll and other personnel routines), finance (e.g. cost accounting) and production (e.g. stock control). A major emphasis for the further development of the company's computing facilities lies in production control applications.

For the following discussion, systems of computer production control will be classified in accordance with three different types of co-ordination.

Co-ordination by standardization, where the activities to be co-ordinated are stable and the rules applied are simple and consistent, means that the need for the computer to react to events is low. The need for fast communication of data between computer and user is therefore also low.

For co-ordination by scheduling, the same high degree of stability and presence of routines can not be expected, and hence the reaction time of the computer is more critical. Structuring and communication of data requires a more advanced computer architecture (e.g. data base).

For co-ordination by mutual adjustment, events are expected to occur randomly. The computer system must react quickly, and particular emphasis is set on communication ability in the computer system architecture. This is also in correspondence with Thompson's view that:

> coordination by mutual adjustment is more costly, involving greater decision and communication burdens, than coordination by plan which in turn is more costly than coordination by standardization.
>
> (Thompson, 1967, p. 57)

Co-ordination here refers to functions and sub-systems within the organization. Computer aided co-ordination by mutual adjustment thus seeks to integrate organizational sub-systems so as to constitute a unified information system within the organization. Fragmented as well as integrated information systems are established through either centralized or decentralized computer systems. Centralized or decentralized solutions refer to where in the organization the computer system resides, whereas fragmented or integrated solutions refer to how much of the organization is directly affected by the system. A computer system may thus be classified with regard to these four categories (see Figure 7.3).

With reference to Figure 7.3, the early computer systems for production control were type I systems where each application occupied its part of the central computer's data processing and storing capacity. There was little or no exchange of information between application modules in the computer. This type of system was based on pre-database techniques (indexed files) with no or only rudimentary capacity for direct man–machine communication. Among type II

integrated system

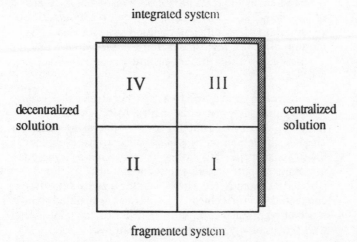

decentralized solution

centralized solution

fragmented system

Figure 7.3 Classification of computer-aided production-control systems

systems are, for example, minicomputers and personal computers, as well as more advanced production-oriented systems such as computerized numerical control (CNC), robots and computer-aided design systems (CAD), all dedicated to a well defined but limited application area (fragmented) as seen from the whole organization. Type III are centralized data base/data communication systems, often modularized in order to incorporate different sub-systems (e.g. purchasing, maintenance) as appropriate. In type IV, networks between decentralized computers and related equipment constitute highly integrated information systems, visualized through, for instance, the CIM concept (Computer Integrated Manufacturing).

Computerization development at NASCO

NASCO's development as a computer user can be described in three phases where the present situation marks the beginning of the second phase.

Phase 1: The development of fragmented applications such as ABC-analysis and MRP on a central computer (IBM). During this phase the (informal) strategy was a further development into an integrated solution, based on COPICS or a similar production control concept. This approach was deferred/cancelled for a number of reasons such as lack of knowledge about COPICS, inadequate and insufficient vendor support, insufficient computer capacity and a systems architecture regarded as too sophisticated to be compatible with the present production situation at NASCO.

Phase 2: A shift from a centralized to a mixed centralized/decentralized approach where the central MRP system (together with other routines for manufacturing planning) is supplemented by a local (decentralized) system for stores control. This stores control system is a module in a minicomputer-based production control system (PROMPT from ICL). The two systems are not hardware connected. The decentralized approach is planned to include also personal computers in order to acquaint users with the company's computer applications.

Phase 3: The long-range computer strategy is an integrated system of the COPICS type for support of the company's production and material control. No official time plan exists as yet for this phase.

With reference to the classification scheme in Figure 7.3 the three phases can be grouped as follows. The main part of the current data processing at NASCO is of type I. There is no central data base for the storing of, for example, production data but each application uses its own data files. The earlier plans of the company were to proceed to type III (COPICS) but are now altered to include types I and II.

However, as we will notice in the following discussion, the different solutions and approaches differ only marginally from each other as far as applicability is concerned, since they are all based on approximately the same conceptual model of industrial production. In the NASCO context, it is this basic model that determines the applicability of the computer-based solutions.

Chapter eight

Materials management at NASCO

Managing the material flow involves two kinds of uncertainty: external uncertainty that stems from an unpredictable environment and internal uncertainty arising from the technical system where functions are interdependent to varying extents. Through boundary units within the supply and stores division NASCO seeks to adjust materials requirements to the constraints and contingencies of the external supply system. Through scheduling and planning of activities in the in-house manufacturing system and of material movements within the company, NASCO seeks to reduce internal uncertainty that affects the material flow.

The material flow consists of parts, components and semi-finished goods that arrive from the suppliers to the goods receiving area, where the flow is split up and sent on in smaller, but still identifiable quantities, to stores and shops within the factory. During the manufacturing process some of the original items lose their identity as they are joined together with other items to form more complex parts and finally end up in the final product, the vehicle. From the product structure description, the bill of material, all individual items can, however, be traced, something that is of vital importance for NASCO's materials management.

The material flow is accompanied by a flow of information (see Figure 8.1) essential for the co-ordination and control of materials supply. This information flow is a necessity for materials control. George Plossl, one of the authorities on materials management, even states that:

> a manufacturing control system is not materials management; its focus is on information, not material.
>
> (Plossl, 1973, p. 9)

As the demand background for NASCO's vehicle production is not particularly highlighted in this study, the starting point of the material flow is arbitrarily set at the production programme specification, stating what NASCO is to produce. This operative target is in the form of a master production schedule that spans over two years and is divided into periods. The master production schedule is based on actual orders, forecasts, strategies which emanate partly from the Ministry of Industry, and on available stock. The master production schedule sets the target for final assembly requirements and is the basis for purchase as well as in-house manufacture of parts. In Figure 8.1 the major streams of the material/information flow are shown.

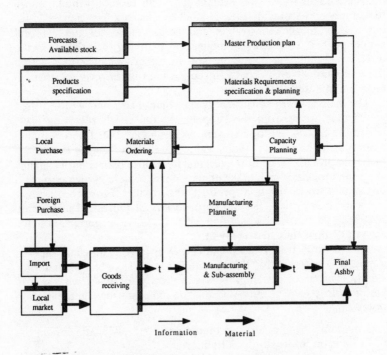

Figure 8.1 NASCO: a schematic view of materials management

Concepts and definitions

The material flow can be (logically) divided into two parts, the materials requirement planning part, where needs are defined and sources of supply are identified, and the materials acquisition part, where materials are ordered and supplied, either from external suppliers or from in-house manufacturing.

The planning of material requirements

The materials requirement planning also has two functions: one where the required parts, sub-assemblies, etc. are identified in relation to their position in the final product structure, and another where the time-span of the requirements are forecast.

The first (qualitative) planning step, calculation of material requirements, is largely independent of production capacity, suppliers' reliability, etc., but reflects the material content of the product in terms of volume and source of origin. The calculation of material requirements for a specific end product does not change from one updating of product data to the next. The result of this step is therefore crucially dependent on the reliability of the stored product data.

The second (quantitative) step, the time setting of material requirements, divides the material needs into time periods as appropriate with respect to lead times, inventory levels and earlier material orders. The calculation of order issuing dates for purchase and for own production is based primarily on the previous step where standard lead times are given for each part.

For materials classified as imported in the parts register, the corresponding need for hard currency is determined and totalled for each period. This is a dynamic phase, as it needs to continually reflect changes in capacity and other resources in order to modify lead time specifications. The tool used for this is a computerized material requirements planning (MRP) system.

Ordering and supply

If planning is the attempt to reduce internal uncertainty through coordination of material needs from interdependent functions, ordering and supply of material is the execution of the plans. External as well as internal uncertainty is involved through purchase from suppliers outside of NASCO and from own production. It must be noted, however, that external material supply is a prerequisite also for in-house manufactured parts.

Logistics theory in general does not consider the function of ordering to be particularly problematic from a materials management point of view (but note well the content of ordering, i.e. to bargain the right price, to select the right supplier, etc.). Instead, focus is concentrated on the supply phase, where concepts like just-in-time have developed. For NASCO, it has been adequate to distinguish between the two phases, as they each have their particular characteristics in the Egyptian environment. A way to characterize the two phases has been to study the lead times associated with the material flow.

Lead times

Lead times are of two kinds: purchase and manufacturing lead times. Reference values for lead times used in material requirements planning are usually kept in the bill of material for each item, and used to determine at what period an order must be released in order to meet the requirement date. For purchase orders, lead times are based on statistical data and suppliers' commitments; for in-house manufactured items, lead times are stored in the routeing specifications.

Lead times are made up of several time elements such as currency remittance, transportation and customs clearance for purchased items, tool set-up and machine time as well as inter-operation time for in-house manufactured items.

Whether lead times for item ordering are realistic or not is determined by the result of the requirements planning. If planning lead time is too short, a purchase order or a shop order is released too late and the number of stock-outs increases. If planning lead time is too long, a shop order is released too early and the result is increased work-in-process with increased planning problems on the shop floor.

Uncertainty, both external and internal, manifests itself as deviations from planning lead times occur. At NASCO this uncertainty appears in both purchase and in-house production, a situation that differs from a traditional Western view as expressed by Smolik:

> Manufacturing lead time is relatively static because it is under internal control whereas purchase lead times vary constantly according to market conditions.
>
> (Smolik, 1983, p. 110)

At NASCO, disturbances influence all three sources of supply.

Description of the material flow

In a flow perspective, materials planning and acquisition can be regarded as a series of activities following upon and related to each other and where each activity (e.g. material supply) starts with an event (material requisition) and ends with another event (receipt reporting), the latter being the starting event for a subsequent activity, and so on.

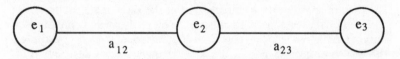

Figure 8.2 Events (e) and activities (a) in the material flow

For this description the material flow can be divided into four phases:

- The calculation of material requirements, i.e.
  the material requirements definition phase
- the scheduling of material requirements, i.e.
  the material requirements planning phase
- the material ordering phase
- the material supply phase

The start and end of the material flow is, of course, set arbitrarily. In this study the start is where the need for materials is established, i.e. at the master production schedule. This implies that any discussions on how needs for heavy vehicles are determined by the company, the

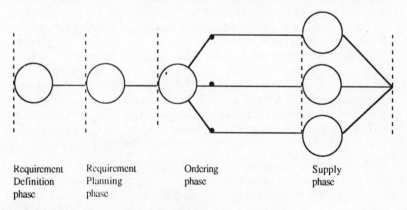

Requirement   Requirement   Ordering   Supply
Definition   Planning   phase   phase
phase   phase

Figure 8.3 The four material flow phases

87

reliability of forecasts, the routines involved in setting the production targets, etc. are all issues which will be left aside, notwithstanding their importance for NASCO. Likewise, the end of the material flow is where parts and materials reach the assembly store. It might seem natural to continue at this point, as materials in the shape of vehicles reach the market, and also thereafter, as there is a flow of spare parts, originally part of the same material flow.

Each phase is assumed to commence by an event and consists of a number of activities, the cumulative duration of which constitutes the lead time for the phase. We will now give a brief description of the four phases.

Material requirements planning

Planning of material requirements is performed with the aid of NASCO's computer-based MRP system. The system is regarded as vital by the production planning management for all material acquisition activities, for external as well as for internal supply. The output of the MRP processing constitutes the basis for initiating purchase and manufacturing orders.

The MRP system at NASCO adheres to a formal technique that is common to all computer-based MRP systems. The technique is briefly presented below.

MRP – an overview

Material requirements planning (MRP) together with order point/ order quantity (OP/OQ) constitute the major techniques for inventory planning and material ordering. Whereas the various OP/OQ techniques control single items independently of other inventory items, and thus with no attention paid to the way components may be used in combination or in assemblies, MRP is based on product structures. The term 'dependent demand' thus reflects a demand for an item resulting from its use in a higher structure level. Therefore, only end products need to be forecast or planned: requirements for lower-level parts and components can be calculated by exploding the scheduled quantity and types of parts needed for the assembly, specified in its bill of material. Sumanth describes the MRP technique:

> The MRP takes the master production schedule for end items and determines the gross quantities of components required by using the product structure records (bill of materials), wherein the gross quantities are obtained by exploding the end item product struc-

ture records into its lower level requirements. Next, by using the inventory status records, the net quantities required are determined by subtracting the available inventory quantities from the gross requirements. Finally, the time periods when the parts should be made available are determined by time phasing (setting back) the lead times for each of them. The MRP produces 'planned order releases' so that purchase orders, work orders, and reschedule notices can be prepared.

(Sumanth, 1984, p. 379)

MRP can now be defined as follows:

MRP is a management planning and control technique. Its initial providing function is to work backward from planned quantities and completion dates for end items on a master production schedule to determine what and when individual parts should be ordered.

(Sumanth, 1984, p. 378)

The MRP technique is based on a number of fundamental assumptions: master production schedules must project a realistic production plan for accurate capacity scheduling. The bill of material needed for product structure explosion must contain relevant data for all items, such as lead times for purchased material, lead times for in-house production, and source of supply. Inventory records must contain accurate on-hand balances for inventory items, etc. It is a general opinion that a major reason for faulty or insufficient MRP systems is inaccurate item data for the bill of material. It is also assumed, at least implicitly, that production conditions such as organizational behaviour, management objectives and fabrication methods do not differ more than marginally among production units for which the MRP technique has been developed. In other words, MRP is based on a model of production activities that are representative for industrial settings of the more advanced industrial nations. A prototype of such an MRP system has been tested at NASCO.

Master production scheduling

The master production schedule is a prerequisite for the planning of material requirements. It is a moving two-year plan that specifies the planned production output of trucks, buses and tractors from NASCO without regard to availability of stock, orders or requisitions. The master production schedule, based on forecasts, contracts and commitments to the planning ministry, is established by a planning committee and shows the planned requirements for each product and all spare parts for the following twenty-monthly production periods. The master production schedule is broken down into assembly parts requirements and thereby controls the acquisition of purchased as well as in-house made parts. The material requirement is here reported as it would be at the start of the final assembly, i.e. no regard is taken to lead times for acquisition and manufacturing.

The requirements definition phase

For the definition of material requirements, information from several sources within NASCO is used:

- the master production schedule provides the time-phased plan for end products.
- the MRP tree (or the group's index file) provides information on the structural design of the products, i.e. what items belong to a product and on which level.
- the parts list provides detailed information about each item.
- the stock file provides information about open orders (i.e. already released orders) and about stock on hand.

The products are specified with regard to their contents in hierarchical structures of six levels. A product is thus built up of product groups, each one containing a number of sub-groups (or sub-assemblies), which in turn consist of parts, the smallest identifiable units. Schematically a structure has the form shown in Figure 8.4.

The heavy vehicle product programme consists of, typically, 10–15 products, around 100 product groups and 40,000–50,000 parts, including spare parts.

The MRP calculation, which uses this product structure to calcu-

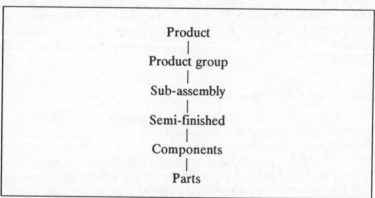

Figure 8.4 The product structure for heavy vehicles

late the number of parts needed to make a product is very demanding of computer capacity. If computing capacity is insufficient, the MRP calculation is not data base oriented but based on 'indexed sequential processing'. This form of data processing does not allow a direct explosion of lower-level items of the product. Instead, requested product groups (level 2) are determined from the production programme (the master schedule for each product) and each production period will reflect the number of product groups requested for that period. This process is repeated until the lowest level of the product hierarchy is reached and all affected items can be summarized per period. Through codes in the parts list, indicating source of supply for each item, items are classified as manufactured, imported or locally purchased. The accumulated needs per period will thus also reflect the need for foreign currency for the import of material.

As noted earlier, the specification of material requirements is a static activity in the sense that with no changes in product structures, a once-established description of the final products can be used repeatedly for specifying materials need. This is certainly not the situation for a production where products undergo constant modification and where new models of products emerge from older ones. At NASCO, being a licence-driven production unit, these changes take place at a low rate and the definition of material requirements can therefore be regarded as a relatively static activity.

The material requirements planning phase

Planning of material requirements leads to the timely release of purchase and manufacturing orders. Item quantities required for different periods, calculated as in the previous step, are compared to stock on hand and to already released orders.

The time-phasing for all items on each level is based on the lead time specifications in the parts list. However, as insufficient information is available for the constant updating of individual lead times, a kind of 'default' lead time is used for requirements planning. The following example can illustrate this: a projected demand of 150 items for each period will, for a stock on hand of 400 items, result in an offset of at least 50 items in period 1 to avoid being out of stock in period 3. The crucial assumption for this, is that the delivery lead time is three periods.

Depending on open orders, the offsetting will or will not result in a net requirement for the specific item in the specific period. The summary of net requirements for all items gives the total requirements per period. These requirements are further processed in the next phase of the material flow: material ordering.

Material ordering

Material acquisition denotes the activity of issuing an order for material. As this order can have three different addresses (local supply, foreign supply and in-house supply) the activities resulting in these orders are partly different and must therefore be treated separately. In the following, however, only a few examples of these activities will be presented. For a more detailed presentation the reader is referred to the more comprehensive study of NASCO.[1]

The activity of issuing an order for import or local purchase is based on a decision by the planning division (material requirements department) from where the request is submitted to the supply and stores division (supply department). The decision is based on available stock and on rules applied by the feeder industry department, which has as its objectives to support national industry and to promote the expansion of this industry in accordance with the industrial development strategy of Egypt.

Local purchase

A requisition for local purchase is submitted to the local purchase department (see Figure 8.5). The requisition is either for a new part or for a re-ordered part. In the case of a new part, the feeder industry department is involved. The feeder industry department, as already noted, is a boundary unit where the involvement of local producers as suppliers to NASCO is jointly discussed by NASCO and representatives from the Ministry of Industrial Planning. For re-ordered parts other routines are followed.

The various activities of this phase, from order requisition to an order placed with a supplier, are of varying duration. Because of this it is not an easy task to predict the duration of the total lead time.

Foreign purchase

The routines for import are indicated in Figure 8.6. A requisition for import of material is submitted to the foreign purchase department by the material requisition department. Foreign purchase is subject to the availability of hard currency in NASCO and to an import licence approved by a purchase committee at the Supreme Council for the Planning of Foreign Trade, a ministerial body.

Using the previous description of events and activities, lead time for foreign purchase can be summarized as follows in Figure 8.7. It must be underlined here that it is not primarily the long lead times that create problems in production, nor the considerable time span

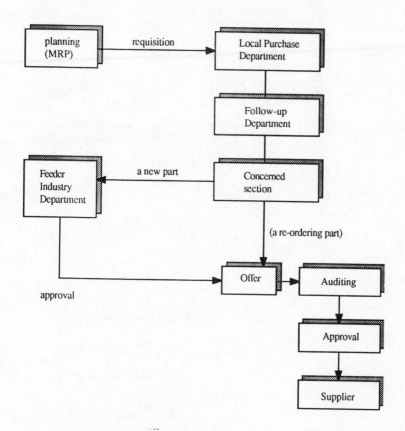

Figure 8.5 Local purchase routines

between planned and actual lead time. These are problems facing most industries, whether in developing or developed countries. What creates problems and difficulties in the planning of production are the great, and largely unpredictable, variations in lead times.

This means that material planners are at times faced with generalized uncertainty, an expression used by Thompson (1967) to denote no or only vague patterns of cause/effect understanding, which are inadequate for recognizing alternatives or establishing control. For the remaining phases of the material flow we will just give the total lead times.

For the ordering of in-house manufactured parts, lead time is

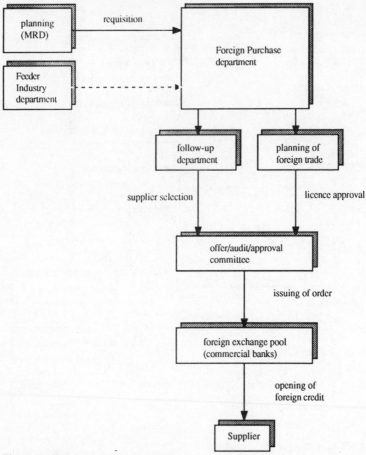

Figure 8.6 Foreign purchase routines

calculated as the time-span when the machine loading department receives a production request until the start of the first operation. This lead time varies from very little to around a year.

For local supply, the lead time starts as goods leave the local supplier and ends as delivery is recorded in the NASCO store's book. This lead time is influenced by a number of unpredictable factors, such as the supplier expediting time, transportation time, inspection and, not least, time for new supply in case of rejection of supplied parts – something that happens, on average, for 10 per cent of inspected material.

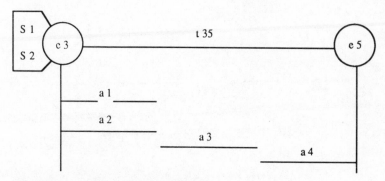

Figure 8.7 Lead time for foreign purchase

S1 Material requisition department
S2 Import order in feeder industry department if a local supplier is not available
e3 Foreign purchase department receives purchase order requisition
e5 Order placed with foreign supplier
a1 Checking of supplier's record ('supplier record committee')
a2 Approval of import licence
a3 Issuing of order. Depending on the previous activity (a2) this activity takes place on average 7 months after the order is received at the foreign purchase department
a4 Activity related to the opening of a bank credit for the transfer of hard currency to the foreign supplier. This activity may be estimated to take 2 months on average
t35 The lead time may vary substantially

For foreign supply, lead time is calculated as in local supply. Unpredictable factors are expediting time at supplier, waiting time for shipping and time for inspection. Average lead time for imported material has been estimated as eighteen months.

Finally, lead time for in-house supply, from the start of first operation until completed work order is recorded in the store, varies significantly due to a variety of factors, primarily unreliable production facilities. The lead time is often much longer than that in the plans, which are often based on standard times.

For operation times, set-up times, tool-changing times, etc., standards are often based on operation manuals of the machine suppliers. The appropriateness of these standard times in the NASCO production environment must be regarded as very limited.

As already noted, there is not only variation in lead times but also a substantial difference in the variations. In Table 8.1 an attempt has been made to assign qualitative measures to characterize the different lead times. Lead-time duration has been characterized as short if

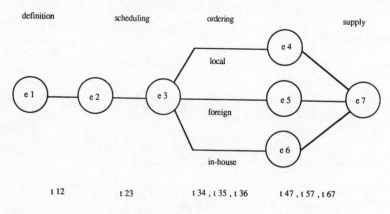

Figure 8.8 Summary of lead times

less than two months, medium if between two and six months and long if more than six months. These figures are arbitrarily chosen. The lead time of the material flow is summarized in Figure 8.8.

Table 8.1 Duration and variance of lead times

phase	Lead-time duration	variance
t12	short	negligible
t23	short	negligible
t34	medium	significant
t35	long	perceptible
t36	long	significant
t47	short/medium	perceptible
t57	long	significant
t67	medium/long	significant

For the variance we use the term 'negligible' to denote no or minimum impact on the lead-time duration. 'Perceptible' indicates that variation in lead time can not be neglected in planning and 'significant' makes planning extraordinarily difficult, primarily because standards essential for planning and control can be based only on highly arbitrary assumptions.

The meaning of variance (from a statistical point of view) here is that the probability that a lead time has a predicted value decreases with increasing variance. Therefore, duration of the lead times t34, t36, t57 and t67 in Table 8.1, although denoted as medium to long, are very indefinite when it comes to expected duration of a specific lead time.

It should furthermore be noted that the variance of the sum of two independent variables equals the sum of the variances. Due to the great number of factors influencing the lead time of each phase, the lead times can be regarded as more or less independent variables. The variance of the total lead time t36 plus t67, i.e. for in-house production, is thus very significant. This, in turn, would indicate that prediction of lead times for in-house production is very difficult.

Capacity planning

Proper capacity planning of NASCO's manufacturing resources is of decisive importance for the material flow, since the decision about source of material supply (import, local supply or in-house fabrication) is partly based on expected available capacity (except for those parts and materials which for cost or quality or other reasons are imported). Misfit between planned and actual capacity therefore has a double impact: it not only leads to in-house production disturbances, but also calls for *ad hoc* solutions, as supply from an alternative source is required.

Input to material requirements planning from capacity planning (as described in Figure 8.1) is based on 'long-term machine-centre loading' (once a year) whereas 'short-term machine-centre loading', executed every second week, modifies already issued work orders.

Long-term machine-centre loading, also called job-shop loading, is thus the basis for capacity planning. In conjunction with the production plan a load calculation is made to show the resulting load of planned orders per machine centre and period during one year. Load in hours is calculated from operation times, i.e. standard times for machine set-up, processing, tear-down and tool-changing. These standard times are often provided by the machine supplier but are not necessarily fully adequate for NASCO's conditions.

The load calculation identifies which machine centres are under- or overloaded. In case of overload, the loading section suggests external supply for the parts in question, whereas underload may lead to additional in-house production.

New shop-loading routines, based on computer simulation, have been developed for NASCO's production. The routines are based on assumptions regarding jobs, machines and times. These routines will be further discussed in Chapter nine.

Resource utilization

Closely related to capacity is the question of machine utilization. Three different sources quoted within the company claim effective machine utilization to be 50 per cent, 50–55 per cent and 37 per cent respectively, the last figure being for a particular month. On average, one machine out of two seems to be working and contributing to the production process. Of a total of 1,386 machines (May 1985), 600–700 tended to be idle for various reasons.

Even though machine stops represent a major cause of production interruptions, other reasons also have an impact on the overall resource utilization. The most frequent reasons are given in Table 8.2. Production stops because of missing material seems to be relatively insignificant (3 per cent of all reasons). It should be noted, however, that 10 per cent of all material requests from stores are rejected due to missing material, which contributes to the fact that more than 90 per cent of all work orders are delayed.

Electricity failures represent on average slightly more than 3 per cent of production down-time, with peaks occasionally reaching considerable heights. In one period, for example, 14 per cent of effective production time was lost due to electrical disturbance.[2] Machine failure, however, is the greatest single cause of production interruptions. There are various reasons for this, although the basic reason is that the machines are old (see Table 8.3).

The age distribution is still more unfavourable according to sources within NASCO, which claim that 'most of the machines are 25 years old and more'. If the present renewal rate of about twenty machines per year persists (1–2 per cent of the total number of machines), the situation will continue to be critical.

Table 8.2 Reasons for production stops

Reason	Stoppage in % of production time	Relative %
Machine breakdown	5	32
Tools missing, defective etc.	4	25
Electricity failures	3	20
Operator absent	3	20
Material missing	0.5	3
Total	15.5	100

Table 8.3 Age distribution of machines.

Machine age (years)	Number of machines	Relative %
0–2	–	–
2–5	70	5
5–10	208	15
>10	1,108	80
Total	1,386	100

With this description of the material flow we will set out to discuss the models that are used to facilitate the cause/effect understanding of production environments.

Part III

Model and reality – a critical view

Aus dem Bild allein ist nicht zu erkennen, ob es wahr oder falsch ist
(It cannot be discovered from the picture alone whether it is true or false.)

Ein a priori wahres Bild gibt es nicht.
(There is no picture which is a priori true.)

Wittgenstein: *Tractatus Logico-Philosophicus*

In this last part we will make use of two things. First, the rather detailed discussion of the production system at NASCO, and in particular the material flow, as an illustration of the kind of problems encountered by production managers in many developing industries. Second, we will make use of the very detailed description of a generalized computer solution for production management, COPICS. To be precise, it is not the description as such, but rather the underlying conception of problems and solutions, of rationality and of organizational behaviour that attracts our interest.

In order to evaluate the applicability and relevance of a Western model in a non-Western environment, a detailed description is needed (i.e. NASCO). The reason is that many of the decisive factors for fit or no fit between model and reality can be observed only on a detailed level. For example, we need to know the age of production machines to understand why production lead times vary, or that local bureaucracy hampers firm orders placed with local suppliers. Without this knowledge about causes and effects we would not be in the position to draw the right conclusions about the applicability of models and, hence, of computer programs based on the models.

The detailed description of COPICS is motivated by its specific approach to production control that is common in the West.

COPICS, being an IBM product, reflects IBM's view on production-management problems and solutions. IBM's view, in this context, can be traced back to a number of individuals in project groups and developing teams. These individuals do not only share value systems and culture patterns, but also the view of industrialization and industrial organization that has developed in the West from the middle of the last century. This view encompasses concepts of rationality, time, values and the like. It is therefore not surprising that practically all production models have the same basic ideas concerning goals and means. In this part we will therefore assess the relevance and applicability of a typical Western production model in Egypt.

Chapter nine

Materials management at NASCO – an analysis

As a point of departure for this chapter we may ask to what extent the material flow of NASCO is successfully managed. This question is of course related to the goals of materials management. When asked to specify the major goals, planning managers on different levels pointed out different goals such as:

- Higher resource utilization
- Cutting of costs
- Optimal mix of products
- Increased productivity

One production executive at NASCO suggested that the major goal is to further increase the local material content in NASCO's products through the expansion of the feeder industry network and, through this, to carry out the government's import substitution strategy.

The definitions of goals and the definitions of successful management of the material flow are related to management level and perspectives. A common definition, that will also be adopted as a criterion for success in this study, is that materials should be available in due time and in the right quantities. As the analysis will take us to different levels, within as well as outside the company, different aspects will be observed where different and even conflicting goals will be encountered.

Even if Egypt's automobile industry which is frequently cited as an example of bad planning cannot be considered inefficient, as stated by Ayubi (1982, p. 367), it is obvious from the case study that if we restrict our view to the material flow it is not efficient in terms of timing and co-ordination with assembly. Whether this is due to bad or inadequate internal planning or to external constraints and contingencies or to a combination of the two will be discussed in the following. For that discussion we need to identify the goals for material supply.

Goals of material supply

An ideal planning situation, where material is to be available at a certain time for assembly, is illustrated in Figure 9.1. The ideal situation would occur if lead times for material supply, which vary depending on type of material and source of supply, were fixed and known beforehand. Release time for a shop order for in-house production or for a purchase order would then be calculated simply by subtracting the actual lead times from time zero, the start of the first assembly operation. This would be a completely deterministic situation. (It is here implicitly assumed that all material is required at one and the same time (t_0) for assembly. This is not quite correct, as material is needed at different times during the assembly process. However, as the assembly process is short compared to parts production the error can be disregarded.)

But even with varying lead times, which is by far the most common situation, starting dates for work orders and purchase orders are specified in a manner similar to the deterministic situation illustrated above, albeit under statistical uncertainty. This uncertainty can be partly compensated with a temporary storage, used as a planning buffer, before assembly starts. Too much variation, however, causes substantial planning difficulties, as illustrated in the following section.

Lead times and variations

The following discussion will be based on a general connection between statistical mean and variance as illustrated in Figure 9.2. The

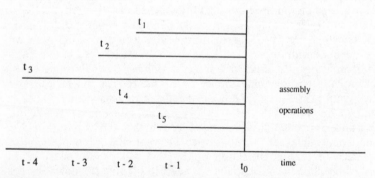

Figure 9.1 Lead times t_1–t_5 of material supply for assembly

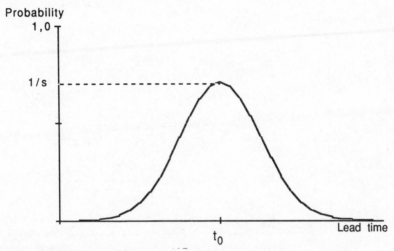

Figure 9.2 Mean value (t_0) and standard deviation(s) of lead time (gauss distribution)

statistically determined lead time has a mean value (or statistically expected value) of t_0, a value that has been estimated, say, from historical data. The variations in lead time can be assumed to statistically follow the gauss distribution, a generalization of the normal distribution. If f(t) denotes the probability that a specific lead time is t, then f(t) is given by the expression

$$f(t) = \frac{1}{s \sqrt{(2\pi)}} \exp\left[-\{(t - t_0) / s \sqrt{2}\}^2 \right]$$

From this expression we note that the probability that a particular lead time equals the reference time or mean value, i.e. $t = t_0$, is given by $1/s \sqrt{(2\pi)}$, where s is the standard deviation. Of relevance for the following discussion is the fact that with increasing variance $s2$, the significance of the mean or reference value decreases. Or, in other words, the greater the variance the smaller is the probability for fit between actual and expected lead time, and hence the higher the degree of unpredictability.

Although this situation is not deterministic as in the ideal case, the uncertainty in lead time is often manageable as long as the statistical distribution of the parameters (t_0, s) can be estimated and, in

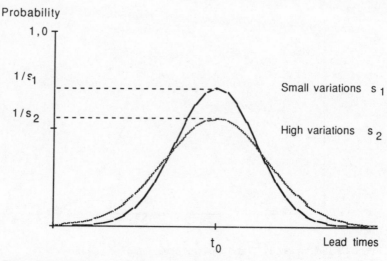

Figure 9.3 Lead times with different variations

particular, s lies within practical limits. Such practical limits are, for instance, a high probability that total lead time lies within the used planning horizon.

Co-ordination and control

Through co-ordination of events and activities along the material flow, NASCO seeks to master uncertainty that arises out of internal interdependency. Three types of co-ordination have been identified, all of them having in common a need for standardized organizational behaviour and an activity structure that is based on the existence of planning standards and reference values. One type of planning standard is lead times.

Table 8.1 illustrated the duration and variance of identified lead times for the different material requisition and supply phases. Table 9.1 summarizes the lead times for each source of supply. The table does not allow any general conclusion as to the ranking of the sources of supply, but it can be noticed that local supply of material has the shortest total lead time and also the smallest variation. It might therefore be tempting to conclude that locally supplied material is more easily planned. The picture does, however, change as we note

Table 9.1 Duration and variation in lead times for material requisition and supply as applied to the three sources of supply (S, M and L indicate short, medium and long respectively)

Source of supply	Lead time	
	duration	*variation*
Local supply	S–M	often significant
Foreign supply	L	often significant
In-house manufacturing	M–L	significant

that 10 per cent of all locally supplied material is deficient in quantity and/or quality, as compared to 2 per cent for imported material.

The high (sometimes very high) statistical variance for lead times means that predicting a lead time is often hazardous, in particular as measured lead times for one and the same item may differ significantly from one time to another (cf. generalized uncertainty). It is therefore not an easy task for the production planner at NASCO to discuss a 'lead-time pattern' for the different items in the bill of material. Co-ordination is therefore hampered by the difficulty in specifying realistic lead-time standards.

Co-ordination is here synonymous with production control as defined by Voris:

> Production control is defined as the task of co-ordinating manufacturing activities in accordance with manufacturing plans so that preconceived schedules can be attained with optimum economy and efficiency.
>
> (Voris, 1966, p. 3)

Managing the material flow, therefore, means exerting control over events and activities in order to provide material in a proper and timely manner. Woodward and Reeves in their study of managerial control point out that:

> every manufacturing business embodies a system for directing and controlling the production task. At its simplest this may be no more than the owner of the business or his representative having decided what he wants to achieve, issuing his orders and making sure that they are obeyed. With increasing size a more elaborate system is necessary and co-ordination becomes a more complex matter.
>
> (Woodward and Reeves, 1970, p. 37)

In the literature on production control, reference is often made to planning either as a separate activity (Eilon, 1962) or as included in the production control itself (Biegel, 1963).

Woodward and Reeves (1970, p. 38) discuss this and the often confusing use of the concept in the literature on production control, and state that 'the confusion arises largely because to control can also mean to direct'. This ambiguous meaning is largely sorted out by Danielsson who points out that:

> for a more extended meaning, the setting of objectives is a vital feature of control whereas for a more narrow meaning, more synonymous with regulation, the objectives are already defined and control is simply to ensure that the objectives are achieved.
>
> (Danielsson, 1983, p. 86)

This concept of control has support also in Woodward and Reeves:

> Precisely defined, control refers solely to the task of ensuring that activities are producing the desired results. Control in this sense is limited to monitoring the outcome of activities, receiving feedback information about this outcome and if necessary taking corrective action.
>
> But planning, setting standards and issuing prescriptions for action, are all prerequisites for control. Without some concept of what should be done it is impossible to make any assessment of what has in fact been done.
>
> (Woodward and Reeves, 1970, p. 38)

As these two modes of control are relevant for further discussion, a brief review of the control concept is appropriate.

The concept of control

From a basic point of view, if we wish to analyse the concept of control, we might start with Webster's Dictionary (1964), where we find the word 'control' derived from the Latin words 'contra' (against) and 'rotulus' (small wheel) with the transferred meaning 'to check against a register'.

The meaning of this is simply to compare the occurrence of an event against some previously decided reference (a value, a combination of values, etc.). How the reference is established or what action can be taken in case of deviation from the reference (and what should be regarded as deviation) is not included in this definition.

An extended meaning of control, also referred to as closed control, involves comparisons and rules for the actions to be taken in cases of deviation from a reference value. The formulation of plans and reference values is, however, outwith this control concept. This is also called cybernetic control.

Goals,
Plans and
formulation
of reference

Comparison → Corrective action

Figure 9.4 A basic meaning of control: comparison

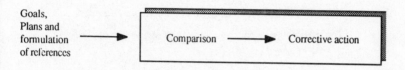

Goals,
Plans and
formulation
of references

Comparison → Corrective action

Figure 9.5 Extended meaning of control: comparison and correction (closed control)

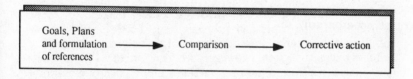

Goals, Plans
and formulation
of references

Comparison → Corrective action

Figure 9.6 Full meaning of control (open control)

When control also involves the setting of objectives and formulation of plans and references, the term open control is used. Open control includes the setting of objectives and formulation of control plans. Reference values (standards) are closely interwoven with goals and plans, and are adapted to changing goals. Open control therefore permits interaction with the environment. Closed control, on the other hand, assumes static goals and plans within a given time-frame and can be considered to be isolated from any influence of the environment in which it exists. The two modes of control are illustrated in Figure 9.7.

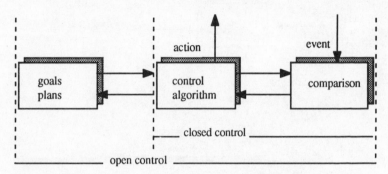

Figure 9.7 The two modes of control: closed and open control

The nature of control

It should be noted here, as has also been emphasized by several writers (e.g. Awad, 1983), that completely closed control does not exist in reality, at least not in organizational behaviour. Even if the control algorithms in Figure 9.7 are supposed to function independently of the goals within a given time-frame, a certain 'leakage' of event-information to the supervisor (planner) must be assumed to occur, and to create a degree of informality in organizational behaviour.

In a technical system, however, we may, in the simplest case (mechanical control), regard interaction with the environment to be practically non-existent. Even if sensors and other technical interfaces change their function due to ageing or material exhaustion, i.e. external influence, thus violating the control strategies imposed by the algorithm, reference values will be reset by the planner (supervisor, authority) rather than left to the control mechanism itself to handle. For a mechanical control system we can thus state that there are no 'own objectives', i.e. the control is truly closed.

In analogy with technical control we will use the terms impersonal and personal control for closed and open managerial control.

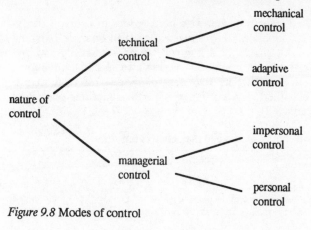

Figure 9.8 Modes of control

Control mode	Technical control	Managerial control
Closed control Open control	mechanical adaptive	impersonal personal

Figure 9.9 Control modes for technical and managerial control

Mechanical control refers primarily to regulation, i.e. strategies and formulations of references are not included in the control tasks but decided elsewhere, often on a higher organizational level. In a mechanical control system all events have equal importance and the only qualitative evaluation of events is the one prescribed by the comparison algorithm.

With reference to Thompson's three types of co-ordination it appears that as the requirements for co-ordination increase from standardization to mutual adjustment, the mode of managerial control shifts from impersonal towards personal control as the task of co-ordination becomes more unpredictable.

111

In this study we need to deal with both modes of control. By expressions like 'production control' we understand open or personal control whereas by 'operational control', an expression that will be used in connection with computer models, we understand impersonal or closed control. The two control modes referred to can therefore be classified in accordance with the functions contained therein (see Table 9.2).

Table 9.2 The two control modes and their control functions

Open/personal control

- set/articulate/define goals
- make a plan in accordance with the goals
- formulate reference values (standards) in accordance with the plan
- monitor
- evaluate performance by comparison with the plan
- make corrections as appropriate

Closed/impersonal control

- monitor
- evaluate performance by comparison with the plan
- make corrections as appropriate

Single and multi-control

The ambiguity discussed above is also due to description level, as closed and autonomous control systems are built into an organization for different purposes and on different levels. Plans on one level have an impact on the control system on a lower level, not necessarily in the routine activities as such but on the formulation of reference values. This can be illustrated in Figure 9.10 where an autonomous control system on one level is 'controlled' via its reference values from a superior control hierarchy.

Different control systems may be interlinked within the same organization, and we use the term multi-control for this as opposed to single control. Multi-control may result in conflicting control criteria with incompatible standards (references). Multi-control occurs as a superior authority 'displaces' the borderline between open and closed control.

Within a closed organizational structure there is either open or closed control, depending on the authority's direct involvement. In an open structure, like NASCO, the authority may be an external

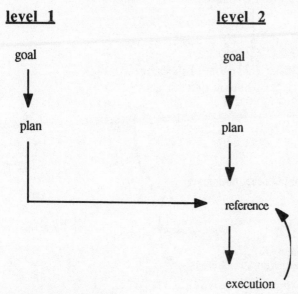

Figure 9.10 Multi-control

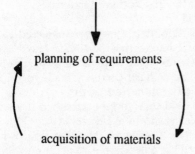

NASCO goals (master production schedule)

planning of requirements

acquisition of materials

Figure 9.11 Control on the NASCO level

Goals (constraints) of government

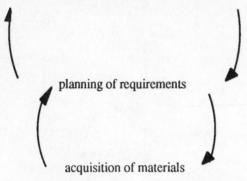

Government Plans: hard currency allocations
NASCO goals: Master production schedule

planning of requirements

acquisition of materials

Figure 9.12 Multi-control: government and NASCO

authority, for instance a government committee, where goals and
plans are determined by the external control structure. The following
example from NASCO illustrates this.

The goals for a production period are specified in the master production schedule on which the material requirements are planned.
The plan is executed through the acquisition of materials. The process is illustrated in Figure 9.11.

Requirements planning for import is, however, subjected to the
allocation of hard currency, which implies that the NASCO goals above
are subjected to external constraints in the form of the (chronic) hard
currency shortage. But this lack of hard currency leads to priority
goals for the concerned ministry (authority) as to investments and
disbursements. Figure 9.10 should therefore be modified if we also
include the government's goals, as shown in Figure 9.12.

Because the currency situation fluctuates, the allocation of hard
currency to NASCO is not fixed but a new application for currency is
submitted for every production period.

With the above image of the external influence of control we will
now analyse the two main streams of the material flow, i.e. purchased
material and material from in-house production.

Co-ordination of the material flow

In the co-ordination of material supply a hierarchical, or rather embedded material control structure can be seen. Within the overall structure of material flow control there is a control structure for manufacturing, within which there is a further control structure for the supply of material to manufacturing operations (see Figure 9.13). Control of material supply for assembly is thus control of purchased material as well as of in-house manufacturing.

The control structure shown in Figure 9.10 illustrates that a plan in one structure becomes the goal in the next embedded structure. For in-house manufacturing, which is part of the core technology that NASCO seeks to protect (or seal off) from external contingency, we have already noted the existence of a set-up of co-ordination methods for shop-order planning and control. But in the lowest/inner control structure of Figure 9.13, where these methods are used for the co-ordination of material requests before shop-order release, the availability of material is dependent on how successful the co-ordination of external material acquisition has been in the higher/outer control structure.

With an allowance of 25 to 75 per cent of requested hard currency per production period, substantial rescheduling of the master production plan is needed for every new period. This leads to changes in materials requisition planning, and hence also in materials supply, with an impact on the availability of materials for in-house manufacturing. Even if internal events and activities in the in-house manufacturing can be co-ordinated, external impacts determine the final result. In the following sections, the control aspects will be analysed in more detail for purchase and in-house manufacturing respectively.

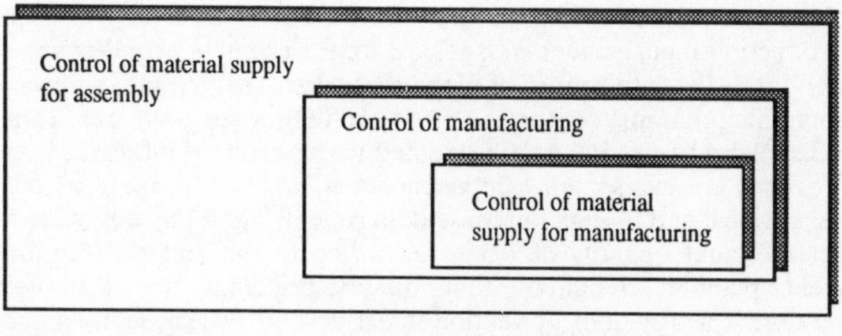

Figure 9.13 Embedded control structures for material supply

Purchased material

Purchase of material means purchase from either local (Egyptian) or foreign suppliers. The various procedures involved in the requisition and supply phases are described in Chapter 8. Here we will analyse those routines which have most impact in this material control context. We will start with the organization of purchasing.

The purchase department

A purchase department has different roles depending on organizational schools of thought. Five or six 'rights' are frequently quoted as the basis for a purchase department's activities: to ensure right material, quality, quantity, time, price and vendor (Stevens, 1985, Corke, 1977). However, as rightly pointed out by Plossl (1973, p. 32), 'the only rights to which the purchase department makes a real contribution are the price and vendor selection' as material, quality, quantity and time are basically the responsibility of the engineering quality control and material planning departments. Selection of appropriate suppliers must then be based on their ability to meet these specifications.

Vendor selection and price negotiation are thus high priority objectives for material purchase. In the light of considerable costs for material (60–70 per cent of turnover, i.e. 180–200 EGP for NASCO), even small savings in percentages result in significant earnings.

For NASCO's materials management the external influence is, however, dominant in the purchase of new or imported parts. 'External influence' means that for local supply, the feeder industry department, which organizationally belongs to the company but functionally is a joint committee with the Ministry of Planning, has the final decision when suppliers are selected. This decision is based primarily on national industrial strategies to promote local industry. 'External influence' can also mean, for foreign supply, that the purchase committee for imports may decide differently from NASCO's production management with regard to, for example, supplier selection and the utilization of approved share of import currency, something that may jeopardize price and delivery negotiations. Table 9.3 shows when suppliers are selected under external influence.

There are two separate departments at NASCO for material purchase: local and foreign purchase department. Their objectives are to get the right quantity of parts, according to the material requirements planning, in suitable time, quality, price and from a suitable vendor. The freedom of vendor selection and pricing is, as noted, rather circumscribed by external factors. In addition, lead times are

Table 9.3 External influence on supplier selection and price negotiation

Purchase of	Foreign suppliers	Local suppliers
old parts	external	–
new parts	external	external

to a great extent influenced by factors outside the control of the purchase departments. Specifying delivery times at order date is therefore of little use. For local purchase, the time needed for the feeder industry department to select an alternative supplier may be up to six months but with considerable variations.

For import, the delivery time requirements are communicated to the supplier. The suppliers are, however, selected partly on delivery reliability conditions, and, being aware of various delays in currency transfers, order approvals, etc., the supplier is reluctant to confirm a shipping date until a firm order is received and the opening of a bank credit is assured. However, as lead times for these activities are in general long and variable, there is no scheduling for imported material until an order-expedition date has been issued by the supplier. But even at that time quoted lead times are exposed to considerable uncertainty as shipping facilities are insufficient.

Purchasing can be seen as a 'boundary function' (as discussed in Chapter four) with the objective to reduce uncertainty that stems from external suppliers. For this purpose the purchase department has its own (open) control system. From the point of view of the materials planning department, however, where plans in the form of material specifications are issued to the purchase department, the control structure of the purchase department is closed. Quantities, qualities, etc. of ordered materials constitute the references in this control structure, and well defined routines specify that remedial actions performed as deviations from references are observed. There is a complete set of formal procedures, based on written reports issued to eight different departments, whenever deviations from reference values (time, quality, etc.) are observed at the goods receiving end.

It should be noted that this local and very formal control structure works technically well and fulfils its objectives. It can be clearly observed that formal routines have a higher status among employees than informal ones, which are regarded as being more primitive. The apparent confidence in formal routines for organizational control which is observable at NASCO will have an impact on the discussion of computerization in the following chapter. A more general aspect

117

of this phenomenon was issued by, among others, Ouchi and Maguire (1975), who point out that the use of impersonal and formal control:

> is largely a result of the demand for quantifiable, simple measures. Paradoxically, output measures are used most where they are least appropriate: in the face of complexity, interdependence and lack of expertise. Under such conditions the manager is suspected of poor performance and, as a result, he goes to great lengths to provide evidence of his contributions to the organization. The organization's need for evaluative information is satisfied, but the quality of that information may be low.

(The term 'output measures' as used in the above extract is synonymous with 'impersonal control'.) Within this closed control structure, arriving goods are controlled according to quality and quantity references. Checks are also made on whether goods arrive on time or not. What is a timely delivery can only be compared to what was estimated at the time of shipment. However, as this delivery time is only a part of the total lead time, i.e. from when the purchase department receives the purchase order until material delivery documents are issued, the closed control structure of the purchase department, with its own goals for delivery times, can not remove the uncertainty in material purchase lead times and the overall objective, i.e. to secure material for assembly operations, is still an open issue.

In the next two sections we will make some further comments on two areas which have already been observed to have an impact on the planning difficulties: the linkages to local suppliers and the shipping facilities for imported material.

Linkages to local suppliers

Difficulties in meeting production targets in terms of delivery dates and quality standards are not an unusual problem in Egyptian industry (see, for instance, Issawi, 1982, p. 150). Difficulties regarding planning and control, similar to those observed at NASCO, are also found elsewhere and the effect is cumulative.

Although delivery lead times from local suppliers, as already noted, are relatively shorter and seem less unpredictable than those of import and own production, this source of supply has essential shortcomings: it has the highest rate of delays and the highest rate of rejections due to low quality, and the price of locally supplied material is often significantly higher than that of corresponding imported goods. The same has been reported from industry in other developing countries, for instance, Tanzania. From a truck manufacturing plant in Tanzania, Brodén reports that:[1]

> as soon as a component was available locally and approved by Scania it was purchased by TAMCO, even if the price for imported components was less. One such example was the locally produced batteries which were three times the price of imported batteries. Furthermore as the local suppliers had monopoly rights for their components and imports were not allowed, TAMCO was forced to keep a large stock of these local components as regular supplies could not be anticipated. The subcontractors were dependent on the Bank of Tanzania for import licenses and foreign exchange allocations for the purchase of raw material. Delays in these procedures was a common problem.
>
> (Brodén, 1983, p. 110)

In his study of linkages between Swedish companies in India and their Indian sub-suppliers, Jansson points out that:

> The greatest problem in buying forging has always been the long delivery times. This is a general problem in India.
>
> (Jansson, 1982, p. 73)

The linkage effects in the automotive industry are generally reckoned as very considerable, and Egypt, having adopted certain restrictions like local content requirements and import restrictions, puts strong emphasis on these effects for her industrial development. Recent studies, however, do reveal significant doubt about these effects, and in a UNIDO report it is even stated that:

> the auto industry as a driving force for industrial growth in developing countries has thus turned out to be a frustrated dream.
>
> (UNIDO, 1984, p. 158)

There is a strong ambition in NASCO to promote and enhance domestic production and, through this, to increase the local material content, in both value and volume, for all the heavy products. The goal has been set at 90 per cent for local content, divided roughly equally between Egyptian suppliers (feeder industry) and own-manufactured parts. The production management at NASCO would, however, prefer a higher share for the feeder industry at a sacrifice to NASCO in order to reduce the pressure on in-house parts production. This attitude has been common among production executives in other Egyptian public industry companies, i.e. to be in favour of an increased share of local content but doubtful about their own company's ability as a contributor to achieve this. The reason is probably the hardship in meeting production objectives under existing constraints.

Shipping facilities for imports

Waiting time for available freight opportunities in foreign ports can be up to half a year, due to the regulations stating that certain freight quotas must be shipped by Egyptian vessels.

An arbitrary survey of vessels entering the Alexandria port in the first half of 1985 (surveyed on ten different occasions) based on the port traffic list regularly appearing in the (daily) *Egyptian Gazette* showed that five ships out of sixty-three entering the Alexandria harbour were of Egyptian nationality. Even if the underlying material for this observation does not allow any far-reaching conclusion, there are statistics from other sources as well (e.g. the *UN Yearbook on International Trade*) indicating that material shipments to NASCO are hampered by the limited share of available transportation capacity.

In-house fabrication

A reduction in the uncertainty related to the in-house made parts is a challenge to NASCO's production management. To a great extent this uncertainty stems from the fact that manufacturing in NASCO faces the same material supply problems as assembly and, in addition, has difficulties with unreliable production equipment. The co-ordination of shop operations, and in particular the prediction of lead times and completion dates, is difficult for a number of reasons such as the discrepancy between standard and actual times and uncertainty of machine availability. This will be covered in more detail below.

Production capacity

As already noted, the decision for in-house production is based (primarily) on expected capacity. The relatively high rate of production disturbances such as machine breakdowns, electricity failures, etc., makes these capacity assumptions hazardous. Even though many machine centres are equipped with similar machines, the total utilization of around 50 per cent means that on average only one out of two machines are available. To this can also be added the reduced flexibility due to missing or inadequate machine tools.

In line productions such as NASCO's where machines are inter-related to each other so that the output of one feeds another, machine availability is particularly important. Interdependency between machines reduces the availability of the line (even though availability of individual machines may be high) and leads to uncertainty in the technical system. A simple example will illustrate this.

In a set-up of, say, n machines, some may be working and some out of order. If q_j denotes the probability that machine number j has a breakdown, then for a total of n machines, the probability that all n machines are working is given by the expression

$$(1-q_1)(1-q_2)(1-q_3)...(1-q_n)$$

For the total probability, Q, that the job flow is interrupted due to machine breakdown somewhere in the chain, the previous expression leads to

$$Q = 1- (1-q_1)(1-q_2)(1-q_3)...(1-q_n)$$

If for simplicity, we assume all probabilities q_j to be equal, say q, we have

$$Q = 1-(1-q)^n$$

If Q is plotted in a diagram as a function of q and for different values of n, we get the following figure.

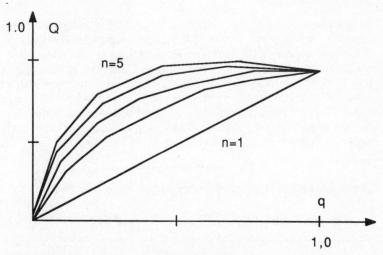

Figure 9.14 The probability (Q) of job interruption as a function of each machine malfunction (q) for n machines

Assuming the breakdown probability to vary continually and uniformly allows us to differentiate Q with respect to q:

$$\frac{dQ}{dq} = Q' = n(1-q)^{n-1} \text{ and thus } Q'(0) = n$$

We can thus conclude that, even for small breakdown probabilities, for each individual machine the growth risk for production-line interruption is directly proportional to the number of machines.

From this simple discussion we can thus conclude that even for reasonably small risks of machine failures the total risk becomes significant as long as jobs can not be re-routed to other machines. For instance, a probability q of 5 per cent (which does not seem unrealistic in comparison to Table 8.2) gives for ten machines a 40 per cent risk for interruptions in the job flow. Expected available capacity as a decision parameter for in-house production therefore seems too simple a criterion as long as this relatively great risk for disturbance is not included. With the current machine maintenance approach, which is primarily *ad hoc*, no significant improvement in machine reliability can be anticipated. A maintenance programme with more focus on preventive maintenance would be an alternative.

Plant maintenance

Maintenance can generally be divided into three categories, each characterized as planned or not planned and repetitive or not repetitive (see Figure 9.15). Preventive maintenance includes minor operations such as adjusting, lubricating and cleaning, but also major jobs such as overhauling. Emergency maintenance involves repair of a machine that is already down or malfunctioning, whereas planned repair involves maintenance related to, for instance, the installation of new machines. In NASCO, emergency maintenance represents around 80 per cent of all maintenance and repair activities. The

Type of maintenance	Characteristics	
Preventive Maintenance	Repetitive	Planned
Emergency Maintenance	Nonrepetitive	Nonplanned
Planned Repair	Nonrepetitive	Planned

Figure 9.15 Characterization of maintenance

figure seems to be representative for other Egyptian industries as well. From a project for improved maintenance at the Helwan Iron and Steel Works in Egypt the report states that:[2]

> It was difficult to implement production plans and schedules to meet the market demands because no one knew when the machine would break or stop. Emergency repairs were 80–90 per cent of maintenance work. The sudden stoppage of one machine may result in the stoppage of a complete production line or even the whole factory.

Cost for plant maintenance has two main components: costs for planned maintenance and costs for unscheduled repairs. A simple expression for maintenance costs is often given by the following relationship where C is the total cost:

$$C = k_1 \times n + k_2\left(\frac{r}{n} - m\right) \text{ if } n < \frac{r}{m}$$

where k_1 and k_2 are costs for planned and non-planned maintenance, respectively, n is the frequency of planned maintenance per machine, r the total annual machine hours and m the time interval within which no maintenance is required. This interval is usually specified by the machine supplier and assumes the use of approved spare parts, regular service and trained service staff. At NASCO, a variety of factors such as ageing equipment and lack of standard spare parts lead to a decreasing time interval m. From the expression of the cost function above we can also notice that the total maintenance costs increase with decreasing time interval m.

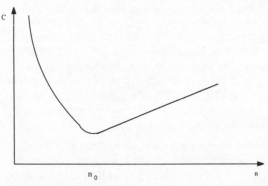

Figure 9.16 Cost of maintenance as a function of maintenance frequency

Differentiating the cost function with respect to n gives cost minimum for

$$n = n_o = \left(r \frac{k_2}{k_1} \right)^{\frac{1}{2}}$$

The cost factor k_1 is basically labour cost and therefore relatively easy to assess; k_2, the cost involved in *ad hoc* repair (machine breakdowns, etc.), can be assumed to be significantly higher than k_1. The expression for n_o above therefore supports a recommendation for a higher maintenance frequency compared to the present situation where maintenance frequency lies very much to the left in Figure 9.16. This is also much in line with the findings from the experiment with improved maintenance at Helwan Iron and Steel.

In-house production of spare parts is a strategy recently introduced at NASCO to overcome the negative effects of currency constraints and to improve the maintenance situation. As indicated above, however, the improvement of machine availability through preventive maintenance could have a significant impact, not only on maintenance costs but also on total production capacity. With a similar example we can assume that a 5 per cent improvement of availability, from 90 to 95 per cent for each of five interrelated machines, would result in an increase in total availability from 59 per cent to 77 per cent, i.e. an increase of over 30 per cent.

Shop floor reporting

Shop floor reporting means, in this context, all kinds of feedback information generated and disseminated as a work order progresses through the factory. Through shop-order documents, concerned departments are informed about the performance of planned activities and the occurrence of unplanned events. Formal feedback through documents and routines ('standardization') has a high standing among planning managers as an important tool to monitor shop floor events and activities. Long reporting lead times and left-out information (e.g. on inspection cards) reduce the usability of the data and increase the risk for decisions based on inadequate information. The strong emphasis on formal routines has been questioned (among others) also by Ayubi:

> then, on all levels of the bureaucracy, the investor cannot but feel the impact of the idolization of papers and documents, signatures and seals, and the complexity, repetition, and overlapping of a large number of formalities and procedures, which of course inevitably lead to various bottlenecks and delays.
>
> (Ayubi, 1982, p. 378)

In parallel to the formal shop-floor reporting system there are informal systems, the existence of which we can observe through the fact that control actions are taken on a short-term basis, e.g. when rush orders are handled or when material shortage is made up through the local hangar (buffer) stores. These informal control structures are fundamental although difficult to identify and articulate.

If feedback is classified according to reporting urgency we can state that the time frames for the formal reporting system qualify only for data collection, i.e. no immediate correction to the production flow is possible. Activities/corrections based on collected data would therefore rather be based on past occurrences. The need for faster feedback to enable faster corrective action is not obvious, as there are not always clear strategies regarding what actions are to be taken. And as long as a standard reference value for a lead time is rather ambiguous, the fast feedback of deviations from this standard is of doubtful value to the planner. A prompt reporting of deviation from the 450 minutes for inspections is not, as an example, particularly relevant as long as total time between planned inspection and the following production step may be weeks, even months.

Improvements under discussion

In this section we will briefly touch upon three production improvements that are currently being implemented by the planning department. The improvements are based on methods and models and will be discussed here simply because to some extent they illustrate the conflict between the planned and the actual situation.

The three improvements are concerned with the loading of machines, the optimal batch sizes for production and a method to reduce the impact of unpredictable lead times. We will briefly discuss the three improvements suggested.

A machine loading system

A mathematical model has been established with the objective of maximizing the probability that order due dates will be met. An urgency factor is calculated which ranks shop orders, which have not yet been released, at a given time with their relative urgencies, giving precedence to those which have the least chance of being completed by due date. This urgency factor is then used to establish a strategy regarding the loading of the machines.

The model (and the results) is based on assumptions regarding shop orders, machines and times. Many of these assumptions are not realistic in the NASCO production context, for example:

- order pre-emption not permitted.
- order cancellation not permitted.
- each machine centre operates independently of the others.
- each machine is continually available for assignment during the schedule period under consideration.

All these assumptions are vital for the model but can not a priori be taken for granted. On the contrary, many of the assumptions, like the availability of machines, are major bottlenecks for the planning situation.

In a probabilistic approach regarding machine loading, based on queueing theory, further general assumptions are made regarding stochastic behaviour of arriving shop orders and order-processing times. Due to the great uncertainty in its basic assumptions, the suggested improvement must be very doubtful.

An economic batch size model

The suggested model calculates total annual costs for processing a number of shop orders, each endowed with characterizations such as set-up costs, inventory holding costs, etc. Optimizing the total aggregated costs for processing the shop orders in a year, with respect to the number of production runs per year, gives an optimal order quantity per run.

Also here the model is based on assumptions which can not be taken for granted. Set-up costs in the routeing documentation, for example, must be very approximate, as production costs are calculated only once per year, and during this period about 50 per cent of all machines have been temporarily out of order. Therefore, the set-up cost per machine per order is a very rough figure, and this uncertainty will greatly reduce the value of the calculated result.

An alternative planning model

Planning for assembly follows roughly the model set out in Figure 9.17 (cf. Figure 9.1). In this model all the different fabrication lead times are established in relation to due dates for assembly operations. However, if one lead time fails (even after uncertainty compensation), assembly operations are delayed.

An alternative approach is being discussed at NASCO, where not the due date but rather the starting date for the longest lead time determines the planning (see Figure 9.18). In concentrating the planning on those parts which have the longest lead times, less attention is needed on the other parts which are likely to be finished in due

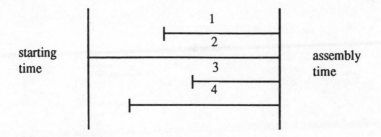

Figure 9.17 Planning for assembly

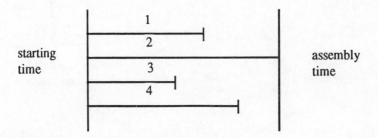

Figure 9.18 An alternative planning approach

time, even ahead of time.

However, as pointed out by Plossl (1973), this situation may easily result in too many work orders being on the shop floor, with increased planning difficulties as a consequence. The effects are best visualized through a vicious circle (see Figure 9.19).

Summing-up

In the analysis of the material flow management at NASCO we have observed the impact from the environment at different levels and in different ways. This impact may be in the form of constraints which the company learns to live with, exemplified through the constant shortage of hard currency that restricts the renewing of production equipment as well as the free choice of material suppliers. We can also regard as constraints the bureaucratic and tiresome formal

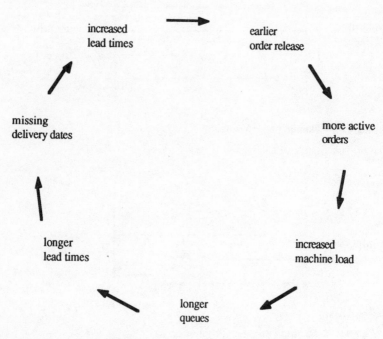

Figure 9.19 A cause/effect pattern for too early order release

system to which the company must adhere for clearance, auditing, etc. of suppliers and supplies.

In order to cope with this external impact, organizational rules and procedures have developed which adapt the company, as much as possible, to these constraints. For the shortage of hard currency, by far the most serious constraint, various organizational steps have been taken that indirectly address this problem. The feeder industry department is one such organizational unit that implicitly seeks to reduce the need for hard currency through fostering and expanding the local supply system.

Impact from the environment may also be in the form of contingencies which the company must face. Contingencies arise largely out of constraints and are severe handicaps for materials management due to the high degree of unpredictability. It has been questioned in the discussion of lead times in Chapter eight, if this unpredictability does not amount to generalized uncertainty where cause/effect patterns are not possible to discuss. Examples of this have been given within the discussion of lead times.

In order to cope with external impact, therefore, there are organizational units ('boundary functions') designed to handle routine transactions as well as *ad hoc* occurrences within their competence frames. Where transactions are more homogenous, as for purchase and receiving of materials, more conventional units were established in the original organization of NASCO. For more specific transactions, like those involved in the promotion of a domestic supplier system, the need for a dedicated function gradually developed and resulted, some years ago, in the feeder industry department. The loosest form of boundary function, from an organizational point of view, are the committees where personnel from NASCO and from outside meet to jointly handle specific tasks, such as auditing of material import orders.

Despite the substantial work within the boundary units and the efforts by planners and managers to protect the work in the factory from undesired external impact, uncertainty persists for the planning and supply of materials. A direct consequence of this is that internal activities such as in-house manufacturing can not be characterized as a closed system.

Uncertainty due to the great variation in lead times, particularly for in-house manufacturing, presents itself differently on different levels.

On one level, uncertainty is caused by interdependence between functions, which NASCO's production management seeks to reduce or eliminate through various co-ordination measures. For example:

1. Machine reliability is directly dependent on the maintenance department's access to skilled personnel, spare parts and other resources. Limited resources (people, spare parts, skills) lead to a pooled interdependence between machine centres, as the use of spare parts for one machine reduces spare parts available for other machines. As maintenance is primarily *ad hoc*, very few maintenance activities can be scheduled.

    A way to improve maintenance might be to introduce a priority scheme where maintenance criteria would decide in what sequence machines should be worked on. This would lead to a sequential interdependence between machine centres, as maintenance of one machine would have to be finished before the next machine could be served.

2. Interdependence between material stores and the shop floor is

sequential, as a shop order can not be released until the material is available. Standardized routines apply, but since the order specifications (quantities, etc.) are different for each order, each material requisition is individually scheduled. Reports on released material are fed back to the stores control department for updating of stores levels, but the lead time for this reporting is not short enough to reflect the actual stores level. This in turn has an impact on the material requirements planning through the balance-on-hand record.

3. Interdependence is reciprocal between inspection, machine centres and the production control department. A product must be processed before it can be inspected and the inspection result may have an impact on the machine centre (too much scrap may indicate that a tool must be replaced). The inspection may also result in changes in a production programme in the production control department. The most appropriate co-ordination here is by mutual adjustment, i.e. fast information is needed for fast action during the production process. However, the information carrier used (the inspection card) is not transferred fast enough to the concerned departments.

On a second level, however, we can observe how internal uncertainty stems from external contingency and constraints. Currency shortage, for example, has a direct influence on co-ordination of internal interdependence. In the three examples given above, this external contingency manifests itself in the first two examples through machines which are too old (with inadequate spare parts) and in material shortages. In the third example, no direct external influence can be observed; the problem is primarily due to inadequate and insufficient working routines. Indirectly, however, inappropriate inspection tools are a result of currency shortage.

External and internal uncertainty are both of significant impact for NASCO's management and we can further categorize these problems into structural and culture-based. Uncertainty can now be ascribed to the organizational structure (shortage of currency) or to cultural phenomena, for example the Egyptian bureaucracy (even if bureaucracy is not necessarily a cultural phenomenon). Figure 9.20 is an attempt to categorize the different forms of uncertainty that we have been discussing in this chapter.

From the discussion about NASCO, its production planning and its management we can now summarize a number of observations regarding NASCO's production with emphasis on the applicability of the computerized control systems. The most important observations are that:

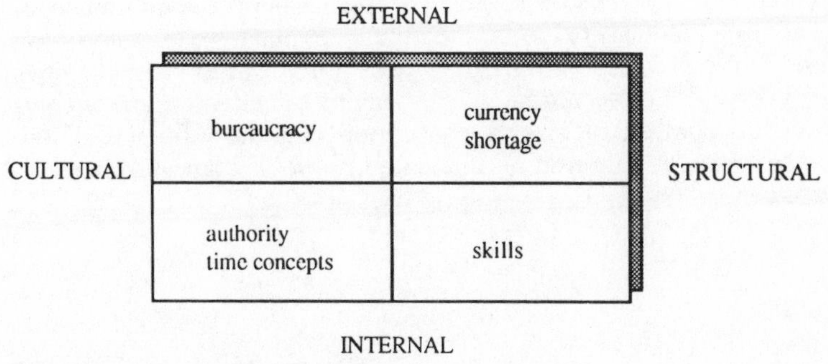

EXTERNAL

burcaucracy	currency shortage
authority time concepts	skills

CULTURAL STRUCTURAL

INTERNAL

Figure 9.20 Uncertainty as a management problem

- production capacity is difficult to predict due to unreliable machinery and other infrastructure constraints such as power failures;
- availability of imported material is difficult to predict due to financial constraints;
- centralized authority hampers taking care of problems as they occur, near the source;
- great fluctuations in lead times make prediction about future events in the production process difficult;
- changes imposed by government in matters relating to the organization or to product programmes interfere with goals and plans of the company.

These phenomena have a direct impact on the preconditions for control and they severely restrict the applicability of the computer systems.

But applicability can here be used in an operational meaning as regards ease of installation and handling, skills and training requirements, costs and benefits, and, not least, end-user involvement. From the preconditions prevailing at NASCO – limited experience, skills and familiarity with computer-based techniques both within and outside the computer department – these operational issues have therefore attracted due attention from NASCO's production management. It should however be noted that the applicability of the production models has been given little, if any, attention.

An explanation for this may be the fact that NASCO management, in accordance with Thompson's (1967) behaviour model, regard the technical core as more or less a closed system where practically all the variables that intervene with the material flow arise internally. It was thus observed that NASCO's production management has a tendency to ignore the fundamental nature of many of the obstacles to production. Instead of regarding them as characteristics of Egypt and Egyptian industry, obstructing the effective use of computerized control, there is a conception that the influence of the obstacles can be removed or eliminated by an appropriate computer system.

Chapter ten

Computerization at NASCO – illusions and achievements

The implementation of a production planning and control system that utilizes the data-storing and processing facilities of a modern computer system has had a high priority at NASCO. The system is intended to improve the co-ordination of the material flow for final assembly of heavy vehicles. Very schematically, the problem to be solved is to determine when, what and how much material must be acquired to meet a given production plan.

The question of what has been achieved and not achieved through the implementation of the computer system needs, however, some further illustrations before we can come to a conclusion. That the computer has contributed to a refined classification of material with respect to usage, costs, etc. (ABC classification) is indisputable. Computer-aided material requirements planning has moreover led to more structural procedures in the material acquisition process, which in turn has pinpointed certain basic organizational problems. It is, however, not evident that the use of computerized control at NASCO has contributed to an improvement of the overall production situation or to improved production efficiency.

The reason for this is that there are factors which are not controllable by the production department and the impact on production exhibits a pattern that is partly not recognizable by the adopted computer programs. Such factors are, for example, the problem of determining standards for planning and the scarcity of foreign currency needed for the purchase of materials and equipment.

In this chapter we will therefore identify a number of 'design features' of the computer programs in order to find out how they match with the production problems at NASCO. As a first step we will briefly resume the earlier discussions about uncertainty and the need for improved co-ordination.

Uncertainty and co-ordination needs

Constrained by external contingencies, NASCO has developed an organizational behaviour that can be partly described within the theory suggested by Thompson (1967). The essence of this theory, as applied in this context, is that external and internal uncertainty partly determines the organization of the company as well as its behaviour. For relations with the environment, from regular input/output transactions to negotiations with suppliers and authorities, NASCO has thus sought to establish boundary functions such as the company's purchasing and feeder industry departments. The tasks of these functions is to build standardized routines to deal with the transactions but also to help cope with unexpected events. As the boundary system incorporates more and more of the in/out transactions between NASCO and its environment and learns, to a certain extent, to cope with external uncertainty, the company's production functions (its technical core), becomes more protected from environmental influence. Uncertainty in the technical core is of an internal nature and basically stems from the interdependency of internal functions. In short, this is a perspective adopted by production mangers and by those who are responsible for developing the new computer-based solutions to production control.

In the *closed* perspective uncertainty is reduced by *adapting* the organization to the environment and by *co-ordinating* production activities. However, co-ordination needs to vary in accordance with the type of interdependence between internal functions so that higher co-ordination needs lead to higher demand on information exchange and communication. This demand is reflected in the design features of the computer programs which emphasize the need for efficient information exchange between the technical core and the different boundary functions as well as within the technical core itself. With different kinds of interdependency between functions, there are also different types of co-ordination needs.

These different co-ordination needs are reflected in various computer concepts for production control. An important observation in this context is that computerized control is here restricted to what was defined as closed control in Figures 9.5 and 9.7. This definition of control implies that goals and plans as well as the formulation of reference values are outwith closed control.

This means that the various computer models for production control implicitly assume that cause/effect understanding is present. Computer control is therefore not suitable to cope with 'generalized uncertainty', a term borrowed from Thompson (1967) to denote that control patterns cannot be established and alternatives cannot be recognized, since no (or only vague) cause/effect understanding is present. But also for uncertainty where cause/effect understanding is present, albeit with a high degree of unpredictability, the number of alternatives may be so high or the control pattern so complex that the control functions of closed control are not adequate.

COPICS – a representative computer solution

The software packages available from different vendors and software houses address production problems either separately or integrated, and with varying levels of ambition. On a relatively high ambition level, the integration of different plant functions into information networks has been technically successful through the advances in computer and communication technology. As already pointed out, however, regardless of the different technical aspects, the computer programs for production control have in principle the same basic logic and assume in their approach to production control a similar organizational behaviour.

In this basic model the emphasis is on the importance of standards for lead times, for operations, for qualities, etc. It is assumed that resources are exchangeable and that priorities can be changed within short time-frames in accordance with replanning of production. For example, that alternative means of transport can be utilized outside the company or that production can be re-routed to alternative machine centres.

It is furthermore assumed that suppliers are exchangeable and that key personnel like purchasers and material planners are knowledgeable about and authorized to negotiate alternative suppliers, materials, prices, etc. In this basic model, uncertainty stems primarily from the product market where competition and free price-setting demand a high degree of flexibility in companies' production planning. The resource and finance markets may be uncertain but not unpredictable. As we have seen, a number of these assumptions do not hold in the NASCO context.

With the basic structure of COPICS as the departure concept, and also as a reference model for production-control systems in general, we will be able to evaluate the computer solutions that have been tested and partly adopted by NASCO. Evaluating the characteristics of COPICS in the light of what we know about the material flow and how it is being managed will then allow us to assess its applicability.

COPICS, focusing on rapid information exchange and immediate reaction to deviations, is real-time oriented and disseminates information, through its integrated approach, between departments. COPICS is aimed at reciprocal interdependence and addresses the need for mutual adjustment between functions and sub-systems, for example between the planning of material requirements and available production capacity.

The idea that employees in different positions will react and take measures as appropriate is fundamental to the integrated communication-oriented systems. The high costs which are often involved for hardware and software to permit interaction with users are thought to be justified by the possibility of further reducing work-in-process and inventories, and by better utilizing available resources through a better command of the information flow.

A crucial factor, however, is the limited distribution of authority in NASCO. COPICS assumes, through the focus on instant reaction, action queues, etc., that users will react. However, middle-level supervisors and managers are not given adequate responsibility for this, something that has also been emphasized by Badran and Hinings (1981) in what they describe as 'lack of authority' in Egyptian public industry. In addition to this, the benefits of using pre-specified measures in the COPICS action queue are also limited by the difficulty of predicting what is a relevant measure in a largely unpredictable environment.

The question of authority has been brought up in other contexts as well. Deyo (1978) thus reports from Singapore that comparison between Chinese and Western-owned factories revealed that the Chinese-owned firms were more centralized and had autocratic, task-oriented authority relationships. The middle-level supervisors were not given adequate institutional authority and responsibility to perform their jobs. The Western firms' decision-making process was, in contrast, more democratic and widely diffused across levels and departments. The integrated and communication-oriented support systems (like COPICS) have a basic design that largely corresponds to a Western decision-making process as far as delegation of authority is concerned.

Organization requirements that can be regarded as characteristic for this alternative are:

- sufficient institutional authority to allow fast reaction in case of deviation from plans;
- standardized routines as a basis for the formulation and specification of appropriate measures to be taken as above;
- reasonably probable estimates of lead times for the setting of ordering periods.

In the following section a number of basic features in the COPICS programs will be compared to the corresponding needs at NASCO. These features are basic in the sense that they are articulated already in the underlying production model. We therefore refer to them as design features.

As we now set out to assess the applicability of COPICS at NASCO, this question of applicability cannot be isolated from whether NASCO's production is regarded as an open or closed system. Applicability must therefore be analysed from two viewpoints:

1. No consideration is taken of external constraints and contingencies but production at NASCO is regarded as a closed sub-system, constrained only by internal uncertainty.
2. Consideration is taken of external constraints and contingencies, such as currency shortage and bureaucracy, which are assumed to contribute directly to the uncertainty facing NASCO's production management. This implies that the open-system perspective is applied to NASCO as a whole as well as to its production functions.

The first approach above (1) can be regarded as rational in the sense that external uncertainty is assumed to be controllable through the boundary functions and that internal uncertainty can be perfectly coordinated. It is furthermore assumed that all relevant variables for decisions are contained in this closed sub-system and as external variables are excluded, the remaining variables vary, but are controllable within the system. We will call this the closed approach. In the second, open, approach (2) we will make use of the findings from earlier chapters, i.e. that the technical core is also directly exposed to external uncertainty.

The following examples will illustrate how applicability must be assessed differently, depending on a closed or an open perspective.

The computer used for planning in general

In the first example we note a feature that is fundamental for all the three alternatives discussed for materials management (namely, that planning implicitly assumes that one of many alternatives can be chosen). In a situation with no constraints this would be true, and even in a constrained situation there may be alternatives. With a closed perspective, external constraints become irrelevant, since they do not, by definition of closeness, have any impact on the material flow. But if the boundary functions of the NASCO organization cannot eliminate or reduce the effects of external contingencies which severely interfere with the production process, then the closed perspective gives an inappropriate image of NASCO. The more we find reason to regard the technical core of NASCO as open, the less relevant, therefore, is this particular design feature.

Design feature

Planning means choosing from alternatives.

1. NASCO as a closed system:
   Re-routeing of the internal production flow in case of machine breakdown is facilitated by the support of up-dated routeing registers and a scheduling system for machine maintenance.
   The selection of alternative supplier's of material is improved with help of a data base of suppliers and a supplier quality-recording system at the goods receiving docks.

2. NASCO as an open system:
   Ageing machines and scarcity of spare parts due to currency shortage result in low machine reliability. Re-routeing is therefore in practice hampered by the high proportion of machines that are out of order.
   Selection of suppliers is restricted by government policy and changing rules outside NASCO. Uncertainty in the allocation of currency and the often very tardy opening of bank credits result in delays in payment and deliveries.

The examples show that the potential benefits of computerization are more obvious for the closed system approach than for the open system approach.

The computer used for manpower planning

The second example focuses on resource utilization and also in this example, adequacy of the design feature can be analysed from either an open or a closed perspective. The design feature here means that manpower can be better utilized through improved machine centre planning.

NASCO as a closed system

Skill files for employees together with an attendance and job reporting system help to match operator and machine centre for particular jobs and operations. Manpower planning is also essential for maintenance, as particular skills required for certain types of equipment must be scheduled in advance.

NASCO as an open system

NASCO as a public company is obliged to receive and employ personnel according to government rules. Improved manpower planning therefore has little effect on the overall utilization of manpower as there is constant and significant over-staffing.

The discrepancy between planned and actual events, for example a specific shop operation, means that detailed planning tends to be inadequate and irrelevant at the actual time. In particular, the most needed and skilled personnel are by necessity used on an *ad hoc* basis.

Skill files for personnel must be frequently updated to reflect the latest status. Earlier notions about formal routines which are hampered by bureaucracy indicate the problem of making such files efficient. The absence of a formal grade system furthermore leads to the recording of relative skills which may be inadequate, in particular for machine maintenance.

The computer used for materials planning

This design feature emphasizes the importance of having standards available for material requirements planning.

NASCO as a closed system

Standards and reference values can be entered and stored in the product data files as well as in machine routeing files for planning purposes. Such standards and reference values pertain to lead times for production and purchase as well as for quantities (for example, economic order quantities) and in principle determine order dates.

NASCO as an open system

Standards and reference values are in practice not possible to specify as the variation in lead times is too high to have any statistical significance that can be used for planning purposes. Most reference values used in the MRP system are therefore inadequate as planning parameters, something that can also be noticed from the great variation in the material flow.

The computer used for capacity planning

The design feature here means that production capacity is controllable within the company and is in practice a question of balancing the work flow.

NASCO as a closed system

The work load is determined by time periods for each machine centre, based on planned order schedules. The result is a prediction of which machine centres will be overloaded and underloaded. Based on this, appropriate measures can be taken to balance capacities and loads.

NASCO as an open system

Capacity shortage is primarily an effect of currency shortage for the import of machines, spare parts, spare tools and repair tools. It is also the effect of inadequate or faulty maintenance procedures or lack of specialized skills.

Uncertainty about available capacity in forthcoming time periods and unpredictability about which machines will be in operation severely reduces the benefit of detailed planning. Production capacity is thus only to a limited extent controllable within NASCO.

The computer used for machine tool planning

The design feature here means that tool availability is ensured through better planning of tool usage

NASCO as a closed system

Tool requirements are scheduled in accordance with planned production. Tool stores can be notified in advance of planned requirements which allows time for tool maintenance and, if needed, for the acquisition of new tools. Tool recall procedures also ensure that tools are returned at the proper time for repair and overhaul, thus improving product quality, reducing scrap and increasing the productive life of tools.

NASCO as an open system

Availability of tools is hampered primarily by a general shortage of tools caused by financial constraints. For the same reasons as mentioned above, planning is hampered by unpredictability which means that tools must be available on an *ad hoc* basis rather than planned long in advance.

NASCO has started in-house production of machine tools and is therefore less exposed to external conditions, provided material and equipment needed for this production is available (which also needs foreign currency and skilled personnel!). The potential for improving tool requirements planning with a computer is substantial. A prerequisite, however, is that routines and procedures be established for the reporting of tool usage.

Computerized control at NASCO – a summary

The examples above show that fundamental design features of the underlying models, some of which have been identified here, can be judged highly relevant or of limited relevance, depending on whether NASCO is regarded as open or closed. We must therefore conclude:

1. that the question of applicability is closely related to the open or closed perspective; and
2. with an open perspective, the computer system as a co-ordination tool has only limited functional benefit for NASCO.

The last suggestion above is in correspondence with findings from earlier research about the applicability of Western organization theories and methods in development contexts. Many successful applications of organization theory have thus been reported because they dealt essentially with the core of the organization and entailed little interaction with its environment. With reference to our adopted terminology of constraints and contingencies we can therefore say that the computer solutions discussed seem to work well when they deal with constraints rather than contingencies, largely because the types of contingencies that have been foreseen in the computer programs, already at the design stage, are mostly irrelevant or inadequate for NASCO.

The explanation for this is to be found in the logic and design of the computer models, where strict assumptions are made about the very fundaments of control, i.e. that goals have been adequately set, that plans have been made in accordance with the goals and that a set of reference values or standards has been formulated as guidelines for execution of the plans. If these functional requirements are not met, the computer systems will fail to perform their tasks for the following reasons:

- the whole control structure in the computer models is based on well-defined goals. If goals are ambiguous or if multi-goals exist, the whole control strategy becomes ambiguous;
- planning implies that different alternatives exist and that a choice can be made. Due to shortage of resources (machines, money) at NASCO there are often no alternatives. Or there is no awareness of alternatives due to lack of communication. Or because of the emphasis on normative Western management

concepts, alternative management concepts, of local character have not been sufficiently considered and elaborated. This reduced range of choice reduces planning to finding a way rather than finding the best way;

● standards are expected to be formulated together and in accordance with plans, and to be based on experience. As, however, accumulated experience of production planning at NASCO does not converge towards a set of standards, due to too much variation in lead times, qualities, etc., there is a tendency for standards and reference values for control purposes to be theoretically formulated. As these standards are the operative parameters that connect the plan with a control algorithm in the computer application program, the performance of the computer system will not be any better than the significance of the standards.

It is thus of vital importance that all the three functions, i.e. goals, plans and standards, not only exist but also are adequate for the actual situation. Otherwise, the computer programs are not able to perform their tasks: to compare actual data with standards and perform according to the algorithm.

Rationality, culture and computer applicability – a remark

In the intersection between Western computer models and NASCO we have had frequent opportunities throughout the study to refer to the concept of rationality as one determinant for organizational behaviour and for the design and development of computer systems. It has been noted that the Western conception of rational behaviour used as basis for the design of the computer systems studied is often inconsistent with the observed conditions and behaviour at NASCO.

The observations at NASCO have been made from an open-system perspective where interaction between NASCO and its environment has been considered important. The question of computer applicability within NASCO can simply not be discussed without considering the external influence on the (internal) production system.

With an alternative approach, i.e. if NASCO had been regarded as a closed organization in which external uncertainty was more or less ignored, a different view of computer applicability could have been expected. As a matter of fact, it has been indicated in the study that the computer models discussed would have exhibited a stronger fit with the NASCO production system in a closed-system approach. Or, put another way, the Western computer models do not cope with an open-system view of NASCO, where an indigenous rational behaviour has developed to cope with events that are unpredictable or indeterminate.

More generally it seems as if the Western culture does not contain concepts for simultaneous thinking about rationality and indeterminateness. The interrelationship between environment and organization, a central theme in this study, has led many researchers to the observation that administrative practice is, in reality, culture bound (e.g. Fleming, 1966 and Luckham, 1971).

This can be seen in a wider perspective as regards Western civilization and its confrontation with other cultures. For example, von Wright (1986, p. 20) raises the question: if each culture has its own conception of reasonableness, whether comparison between different forms of rationality is therefore at all possible. And Bärmark (1984, p. 6) claims that criteria for rationality and knowledge cannot be formulated unless they are tied to the social and historical context.

The examples of rational thinking that we have observed in the computer models of this study, for example in the assumptions about authority and predictability, belong to a Western rationality as discussed earlier. The applicability of these models in a non-Western society can therefore not a priori be expected unless the same rationality is at hand. With reference to von Wright (1986) and Bärmark (1984) above, different cultures have different sets of rationality. Applicability of computer models, or degree of fit between model and reality, can therefore be assessed only through analysis on the level and in the place where the model is to be applied. Further conclusions of the study are therefore that:

- a computer model is designed and developed within the framework of a given cultural conception which includes norms of rationality that determine the model design;
- in another cultural setting, where different norms of rationality apply and influence the behaviour of an organization, the applicability of the model can only be determined in direct confrontation with its actual application in the organization;
- in another cultural setting, where different norms of rationality apply, an organization may be looked upon as a closed system. If the boundary between the environment and this closed system is suitably drawn, then any model and norms of rationality can be made to fit;
- in the case of NASCO, where the open-system approach has been used to study the organization's interaction with its environment, the conditions and the problems facing the company's production management exhibit significant

143

discrepancies in comparison with those of the computer models. The difficulty in predicting lead times, just to take one example, turns out to have a major hampering effect on the usability of the computer systems studied;

- we can therefore state that the computer system does not have the technical characteristics or the technical capability that in the first place determines its applicability. Of primary importance seems to be how the problem-solving design of the program, and the implicit assumptions regarding rational behaviour and practices can be utilized.

With this we will leave NASCO and the detailed discussion about the production environment and the applicability of a practical kind of computer software (namely for production control). In the last chapter we will again take a wider and more general approach and attempt some more general conclusions regarding models and reality.

Chapter eleven

The NASCO analysis in perspective

For the analysis of NASCO we have used a description model that is based on contingency theory as presented in Chapter three. Within this theoretical framework we have described how NASCO is exposed to external uncertainty, which also affects the internal production system. But in the discussion about applicability of Western models in non-Western contexts we must, of course, critically examine how adequate the adopted description model is for the analysis of NASCO and its production organization.

The contingency theory – some comments

The contingency theory can be traced back to the 1960s and the pioneering research of Burns and Stalker (1961), Lawrence and Lorsch (1967), Thompson (1967) and Woodward (1965), who concluded from their applied organizational research that organization structures tend to not only be based on the tasks assigned to the organization but also on other factors such as environmental conditions, technology, culture patterns, etc. The core concept of the theory is openness; in other words, that organizations must be conceived as open systems and that it is external uncertainty that constitutes this openness.

Thompson, frequently referred to in this study, found that this openness, as opposed to an earlier dominating conception of organizations as closed systems, becomes apparent once the organization is regarded *together* with its environment. With this approach, Thompson suggested that organizations respond to external and internal uncertainty by different forms of rational behaviour, adjusted to cope with different forms of constraints and contingency. For the analysis of NASCO we must ask to what extent this rational behaviour is characteristic also for NASCO in general and whether NASCO's behaviour in materials management, in particular, can be explained by the contingency theory.

The contingency theory in development contexts

It is obvious from the previous chapters that, organizationally, NASCO behaves largely in accordance with Thompson's 'norms of rationality'. As Thompson suggests that organizations under norms of rationality seek to minimize the power of environmental elements over them by maintaining alternatives, we can directly identify the feeder industry department of NASCO as an example of a joint venture with local suppliers in order to reduce the kind of uncertainty that comes from currency constraints.

The theory used to describe the behaviour of NASCO should therefore include ways to incorporate the environment and its influence. The contingency theory qualifies for this. But is the contingency theory sufficient? Does the contingency theory distinguish, for example, between uncertainty from currency constraints and uncertainty that stems from bureaucratic constraints? And is it the same kind of rationality that the NASCO management exerts to reduce external uncertainty in the two situations?

Figure 9.20 in Chapter nine is an attempt at categorizing different forms of uncertainty into external or internal and cultural ('informal') or structural ('formal'). The contingency theory recognizes these categories. But we must assume that uncertainty that stems from cultural behaviour and local influence is not adequately described and explained in a general model or theory. Different forms of uncertainty give rise to different forms of rational behaviour which in turn give rise to different measures. These measures thus express a rational behaviour within a given value system. The adequacy of the measures can therefore be assessed only within that particular value system.

Norms of rationality in complex organizations may be at odds with the basic values and orientations of developing countries, where culture does not typically incorporate the skills required for the operation of complicated technology. In those societies the homogenizing influence of culture therefore acts as a constraint on organizations, as factors retarding organizational action under Western rationality norms.

Thompson notes in this context that the homogenizational influence of culture unmistakably appears also in modern societies when complex organizations doing basically similar things are studied and compared. Culture thus acts as a constraint in modern societies as well as in societies in transition (e.g. Egypt), as new technologies call for talents which only gradually become incorporated into the cultures. Some further examples of uncertainty that stems from cultural influence are:

- different concepts of time (Niehoff, 1959; Hall, 1959);
- authority of the elder (Pizam and Reichel, 1977);
- collective responsibility (Harris and Kearny, 1963).

In an attempt to relate cultural influence to specific organizational concepts such as co-ordination and control (e.g. materials control at NASCO), Caiden and Wildavsky point out that:

> Rational decision-making and rational models of planning, scheduling and operations control require precise time phasing, speedy reporting and exact knowledge. The lack of these factors and the poor information base in poor countries make rationality difficult to achieve.
>
> (Caiden and Wildavsky, 1974, p. 13)

Different perspectives, however, give a different signification of rationality. Rationality, or lack of rationality, on one level may be regarded vice versa on another level, and rationality from one perspective may look different from another. An Egyptian public enterprise, for example, is in addition to its role as manufacturer also shaped to be part of the process of political and economic change and development. What is rational here may be different in the private enterprise where political issues are of marginal importance.

Badran and Hinings (1981), in their study of Egyptian public enterprises, thus raise the question to what extent organizations in a public sector enterprise in Egypt exhibit the same organizational characteristics as those in developed countries. The study revealed a major difference in the level of structuring of control and in the distribution of formal control. Youssef states in this context that:

> the social system in Egypt influences managerial values which in turn influence behavior. Moreover, the economic system imposes constraints on that behavior in terms of available resources and the institutional framework of society.... The impact of cultural values on these theories is equally significant.
>
> (Youssef, 1979, p. 367)

Rationality is, as we have earlier argued, a contextually related concept, and, in the study of organizations in different cultures, different needs arise as to what characterizes rationality and rational behaviour within a particular culture. This motivates the treatment of culture as an independent environmental variable that influences the appropriateness of different organizational forms. Pondy and Mitrof

(1979) have thus advocated that organization theory move 'beyond open system models of organization' to a 'cultural model' – a model that would be concerned with the higher mental functions of human behaviour, such as 'language and the creation of meaning'. Smircich, in discussing concepts of culture and organizational analysis, summarizes that:

> it is apparent that the open system analogy continues to be a dominant mode of thought in organization studies, but that now the idea of culture has been incorporated and given prominence as an internal variable as well as an environment variable.
>
> (Smircich, 1983, p. 355)

In the last section we will give some examples of management behaviour at NASCO that illustrate why a non-Egyptian model is not complete as a description model of this management behaviour.

Some examples of management behaviour

Badran and Hinings (1981) report on the concentration of authority from their study of thirty-one Egyptian public enterprises. As they separated centralization (defined as locus of authority within the organization) from lack of autonomy (defined as the extent to which the locus of authority is outside the organization's own hierarchical structure), the result of the study supported the hypothesis that Egyptian public enterprises exhibit lack of autonomy and centralization of authority.

This is reflected through the frequent reference to 'committees' as joint-authority bodies, the loci for decisions. From the NASCO survey there are various examples of this: external suppliers are selected through a supplier's record committee; changes in manufacturing plans are (partly) under the auspices of the planning committee, etc. (A planning committee is an assembly of individuals from NASCO and an external body, e.g. government.)

In analogy to the committee concept there is the reference to 'departments' as loci of authority within the organization when no external influence is present. 'Committees' and 'departments' as authority concepts thus seem to be used as collective decision makers when issues like 'who decides' are brought up. One explanation for this may be found in the scarcity of qualified managers, a severe problem in Egyptian industry as noted by non-Egyptian (Issawi, 1982) as well as Egyptian writers (Tibi, 1984). Due to this scarcity it is difficult to find one individual with sufficiently broad managerial skill to qualify for high-level management, particularly in the bigger enterprises. A group of individuals, a committee, will therefore share the responsibility.

Youssef (1979) suggests, in this context, that the reluctance to be responsible for a decision in the hierarchical Egyptian society is a further reason to share responsibility among individuals.

As pointed out by different writers, e.g. Streeten (1981), Stewart (1978) and Youssef (1979), Western management systems are regarded as reference systems by developing countries, and Western research seems to be the main basis for management textbooks and for training in these countries. Doubts about Western organizational concepts and their applicability to NASCO are shared more by senior managers than by lower-level managers, who do not seem to be particularly bothered by the weak fit of organizational theory as illustrated by, for instance, Badran and Hinings (1981).

We gave examples of this in the three briefly described measures suggested to improve co-ordination and planning of NASCO's production. As noted, the theoretical requirements imposed by the models had limited correspondence with the actual situation at NASCO. The researchers who suggested the adoption of the models claimed substantial benefits for NASCO, whereas senior managers showed considerable doubt about the usefulness of the theories at NASCO.

This hints at another difficulty for the gaining of experience and for management training in the Egyptian socio-economic system: the very weak bonds and lack of mutual understanding between industry and the academic establishment. This has been emphasized by several writers. Moore, for example, reports that:

> applied scientists privately observed that public sector managers had no interest in carrying out research to improve products that they could sell anyway on the protected market. In response, informants who were managers claimed that the research conducted in the universities was too theoretical and removed from industrial problems. A former minister who had bridged the worlds of academy and industry opposed much of the professors' research in the engineering sciences as merely duplicating work done elsewhere.
>
> (Moore, 1980, p. 93)

Moore also reports from the Egyptian National Research Centre (NRC), with roots in the industrial community, that:

> its scientists usually consider industry's research inquiries ridiculous and unworthy of their attention.
>
> (Moore, 1980, p. 93)

A similar standpoint was recently heard from a NRC-staff member, claiming that:

> the bonds between R&D and industry are very weak since R&D must work with advanced problems to keep pace with international R&D whereas industry is oriented towards local markets with limited needs in quality requirements.

As a consequence of the weak linkage between university and industry, students (who constitute the recruitment base for tomorrow's industry managers) become more familiar with Western organization models and management concepts than with actual local industry behaviour. Managers in general, and in particular those at middle-management level, are therefore ill-prepared to tackle indigenous problems in local industry, and there is a tendency to over-value technical solutions to organization problems and to adopt, uncritically, Western management concepts and models.

The following example shows, on a micro level, management behaviour in eight departments of two Egyptian enterprises in comparison with a Western pattern. Even though the material is insufficient for an extensive analysis, some conclusions may, with due caution, be drawn from the statistics on employees, managers and technical office equipment (typewriters, copy machines, etc.).

Woodward (1965) observed a linear relationship between technology, here used to mean office equipment, and the ratio of employees per manager. One rational explanation could be that an increasing number of employees per manager leads to more administrative work which in turn requires more office equipment.

Table 11.1 Employees, managers and office machines in eight departments at NASCO and El Mehalla el Kobra Spinning and Weaving Company

Employees	Managers	Employees/ manager	Office machines
100	26	3.8	20
25	1	25	1
30	6	5	1
25	1	25	0
78	5	15.6	3
147	7	21	3
14	1	14	1
30	3	10	3

Table 11.1 does not support Woodward's observation as regards these eight departments. On the contrary, there is a relatively strong

negative correlation (–0.79), indicating a decrease in units as the ratio of employees per manager grows. Instead there is a very strong correlation (+ 0.97) between number of managers and office equipment, considerably stronger than between the number of employees and office equipment (+ 0.56). It seems therefore that it is not the size of the department that determines the volume of office equipment but rather the number of managers. Or, phrased differently, technology seems to be related less to the organizational structure and more to the management role. This is an example of behaviour which has no explanation in a Western rationality model. Instead, we probably need to extend the theory frame to include cultural influence as well.

From the discussion in this chapter we conclude that the contingency theory helps to structure the analysis of NASCO and to provide a guideline in understanding cause/effect relations. As the contingency theory, however, is based on models (e.g. of a production organization, cf. Figure 4.1), but also contributes to the creation of models (e.g. relations between a company and its environment, cf. Figure 4.2), we need to enter a higher level, as we did in Part Two of this book, and again discuss models and their applicability from a more general perspective.

Some concluding reflections

Das Bild stimmt mit der Wirklichkeit überein oder nicht; es ist richtig oder unrichtig, wahr oder falsch.
(The picture agrees with reality or not; it is right or wrong, true or false.)

Das Bild stellt dar was es darstellt, unabhängig von seiner Wahr- oder Falschheit, durch die Form der Abbildung.
(The picture represents what it represents, independently of its truth or falsehood, through the form of representation.)

Wittgenstein: *Tractatus Logico-Philosophicus*

Chapter twelve

Computerization in developing countries – model and reality

Let us return, in this last chapter, to the basic question raised in the preface to this book, i.e. what meaning do models have outwith their original cultural context? As computer programs are based on the models it seems inevitable that questions about applicability should be seen in this broader perspective.

Models are images of reality in such a way that facts of reality are interpreted, through the language, into a structure of words which must have something in common with the structure of the facts. This latter structure must obviously be known a priori in order to structure the words into a meaningful model. This structuring, we argued, was culturally affected in the sense that words of the model, explanations, symbols, etc. must allude to something that belongs to the cultural environment in order to be intelligible and to make sense. The question is, then: what meaning has the model in another cultural context? How can such a model help to create understanding of, for example, causes and effects in a particular situation in a country like Egypt if there is a different cause/effect structure in the model?

The explanation value of many models is questioned also in the West, in particular as simple and lawful relations in economic, technical and social systems – where individuals interact with technology – which can easily lead to complex, hazardous and even completely unpredictable situations. In systems with three or more parameters, which is not at all unusual even for simple systems, the combination of parameters may, under certain conditions, lead to chaotic situations.[1] Instead of accepting that a model is not completely reliable unless it equals reality in complexity, and hence is of no use as a model, the desire to make reality intelligible and to predict the future leads to over- simplification in our attempts to interpret reality. Sometimes we therefore react *as if* all necessary facts had been transformed into the model. This could be noticed from the NASCO study: the computer system provides information for inventory control *as if* materials were supplied in accordance with the underlying model; or the computer system provides necessary information for production planning *as if* machines were not out of order due to missing spare parts.

The deterministic and mechanical world view that has dominated Western thinking and which has been so important for the development of, for example, computers, has emerged within the scientific (and philosophical) framework that spans from Newton to Einstein. However, a new awareness seems to have emerged which questions many of the existing rationality concepts and which accepts that interpretation of reality is more complex than has been assumed. The fact that most processes of dynamic, economic, technical, social and biological systems are non-linear creates a need for partly new thinking in models.

This need for new thinking also applies, of course, to developing countries, where low utilization of computers and uncertainty about their contribution to the development process should lead to discussions about alternative models rather than to new techniques. In the South, however, the misfit between models and reality is not only because reality is wrongly interpreted but also, as we have seen, because other meanings of rationality apply. For example, better utilization of capital was a driving force for computerization in Western industries in the last half of the 1970s and 1980s. In many developing industries, however, protection of industry makes domestic production monopolistic, or at least nearly monopolistic, which introduces inefficiency, both directly because monopolists exploit their market strength by producing less and charging more, and indirectly because lack of competition reduces the incentive to keep costs low. Even if protection fails to increase monopoly power, sheltered domestic producers will have less incentive to pay attention to quality control and innovations. Under these conditions there is no real incentive to invest in computers for improved quality control or better cost efficiency.

From our study of Egypt we noticed how efforts are made to evade (but also to exploit) the effects of protection and monopolism. The same applies to other developing countries and it has been estimated that a not insignificant share of gross domestic product (GDP), in those countries with protected industry, is spent on exploiting or evading the distortions caused by protection.[2]

Other conditions thus lead to another organizational behaviour and to different rationality concepts. The way to more effective computer usage must start here, with the development of models that reflect local needs and conditions. This development must be based on reality rather than on foreign models as long as the foreign models do not correspond with the actual reality and, in particular, as reality tends to be set aside as soon as a formal theory is adopted.

Notes

2 Computers in developing countries

1. Developing countries denote here the first three groups in the UN terminology classifying the world's nations into five groupings, namely low-income and middle-income economies, high-income oil exporters, industrial market economies and non-market industrial economies (The World Bank).
2. A very informative article about rhetoric and computers, with striking examples, can be found in Rahim, S. and Pennings, A. (1987) 'The Rhetoric of Computerization'.
3. Source: International Data Corporation (USA) and l'Information de la Societé (France).
4. An IBM executive in Tunis thus stated that due to the small Tunisian market it is useless even to file a Request for Price Quotation (RPQ). At the IBM laboratory in Sweden, responsible for systems software development and related RPQ support on the international market, no RPQ from a developing region had been filed, at least not for the last six years (interview).
5. Kaul, M. and Kwong, H.C. (eds.) (1988) *Information Technology in Government – The African Experiences*.
6. See, for example, Naranyan *et al.* (1981).
7. Extracted from a report by Eloranta, E. (1983).

3 Model and reality – a conceptual discussion

1. Blake, J. (1985) 'Accountants and the Finance Functions' in Elliot, K. and Lawrence, P. *Introducing Management*.
2. This is further discussed in, for example, Janik, A. and Toulmin, S. *Wittgenstein's Vienna*.
3. Kuhn, T. (1962) *The Structure of Scientific Revolutions*.
4. A rich book on chaos theory and its consequences is Pagels, H. *The Dreams of Reason*.

4 Industrial production as a model

1. This law of Requisite Variety can be formulated as 'only variety can absorb variety'.

5 Industry in Egypt

1. The four different exchange rates are (May 1987): the official fixed rate of 0.7 EGP to the dollar, set in 1977, used for Suez Canal tariffs, for oil and cotton export and for the import of basic commodities. There is also an official commercial bank rate, 1.36 EGP to the dollar, for customs and similar transactions, and a free market bank rate, 2.15 EGP to the dollar. Finally there is an unofficial free market rate that varies but which is often about 20 per cent above the free market bank rate.
2. Different sources of information tend to give different and partly inconsistent figures. An example is given in Lind (1988) where a comparison of statistical data from different official sources for Egyptian exports and imports in 1980 vary; for imports, from 4,418 to 6,814 millions of US$, i.e. a variation of 50%.
3. *South*, April 1986 and in a report from the US Embassy in Cairo, September 1982.
4. *South*, December 1985.
5. *South*, December 1985.
6. Report from the Swedish Embassy in Cairo, 1984.
7. Girgis (1977) shows that imported inputs as a proportion of total output rose from 6.8 to 16.4 per cent between 1954 and 1970.
8. Four countries – Argentina, Brazil, China and India – have accounted for almost all production by the developing countries since 1966 (Source: UNIDO ID/304 1983).
9. See, for example, Frank (1978).

6 NASCO – a company presentation

1. Interview with director of planning, NASCO.
2. *Le Journal d'Egypte* (Fevrier 15, 1985) 'Le production envisagee est de l'ordre de 3000 camions lourdes et de 600 autobus par an'.
3. *General Motors Newsletter* of June 17, 1986, Detroit.

7 Vehicle production at NASCO

1. According to a NASCO production executive, a locally supplied part that costs up to 15 per cent more than the corresponding imported part is still preferable.
2. ABECOR report, April 1985.

8 Material management at NASCO

1. See Lind, P. (1988) *On the Applicability of Computerization Production Control in an Egyptian Industry.*
2. Ayubi (1982) 'Power failures have indeed become a serious problem for many industries, and the obtaining of premises for industry and business is normally very difficult and in most cases extremely costly'.

9 Materials management at NASCO – an analysis

1. TAMCO: The Tanzanian Automobile Manufacturing Company, Scania: SAAB-SCANIA of Sweden.
2. Tantawy, D., Hamdi, A. and Kader, S. (1984) *Experienced Gained Through the Implementation of Maintenance Management Project in the Egyptian Iron and Steel Industry.*

12 Computerization in developing countries

1. See Pagels, H.C. (1988) *The Dreams of Reason.*
2. For example, the 1979 GDP figures for some selected countries are as follows: Turkey 5–10 per cent; for Pakistan 6 per cent; for Brazil 7 per cent; for Mexico 3 per cent and for the Philippines 4 per cent (see *The Economist*, September 1989).

Bibliography

Periodicals

Administrative Science Quarterly, New York (quarterly)
Business India, New Delhi (weekly)
Business Week, New York (weekly)
Datamation, New York (monthly)
The Economist, London (weekly)
L'Egypte Contemporaine, Le Caire (quarterly)
Ekonomisk Debatt, Stockholm (8 issues per year)
Entwicklung und Zusammenarbeit, Bonn (monthly)
Information Technology for Development, Oxford (quarterly)
Le Journal d'Egypte, Le Caire (daily)
Organization Studies, New York (quarterly)
South, London (monthly)

Non-periodicals

Abdel-Khalek, G. (1982) 'The open door economic policy in Egypt: its contribution to investments and its equity implications', in M. Karr and S. Yassin (eds) *Rich and Poor States in the Middle East*, Cairo: The American University in Cairo Press.

ABECOR Country Reports, London: Barclays Bank.

Amin, G.A. (1982) 'External factors in the reorientation of Egypt's economic policy', in M. Karr and S. Yassin (eds) *Rich and Poor States in the Middle East*, Cairo: The American University in Cairo Press.

—— (1983) 'Economic and cultural dependence', in T. Asad and R. Owen (eds) *Sociology of Developing Countries: The Middle East*, London: Macmillan Press.

Ashby, W.R. (1956) *An Introduction to Cybernetics*, London: Chapman & Hall.

Automotive Industry in Egypt (1985) El NASR Automotive Manufacturing Co. (information brochure), Cairo: Al-Ahram Commercial Press.

Awad, E. (1983) *Systems Analysis and Design*, Homewood, Ill.: Irwin Inc.

Ayubi, N. (1982) 'Implementation capability and political feasibility of the open door policy in Egypt', in M. Karr and S. Yassin (eds) *Rich and Poor States in the Middle East*, Cairo: The American University in Cairo Press.

Badran, M. and Hinings, B. (1981) *Strategies of Administrative Control and Contextual Constraints in a Less-Developed Country: The Case of Egyptian Public Enterprise*, Organization Studies, 2.

Bairoch, P. (1975) *The Economic Development of the Third World Since 1900*, London: Methuen.

Bärmark, J. (1984) *Kultur, Medvetande, Paradigm* (in Swedish), Göteborg: Institut för Vetenskapsteori, Götenborgs Universitet.

Benmokhtar, B. (1984) 'Regional computer cooperation and socio-economic development, case of Morocco', in R. Kalman (ed.) *Regional Computer Cooperation in Developing Countries*, Amsterdam: North Holland Publications.

Biegel, J. (1963) *Production Control*, Engelwood Cliffs, NJ: Prentice Hall.

Blake, J. (1985) 'Accountants and the finance funtions', in K. Elliot and P. Lawrence (eds) *Introducing Management*, London: Penguin.

Bloomfield, G. (1978) *The World Automotive Industry*, London: David & Charles.

Brodén, M. (1983) *From Transfer to Acquisition of Technology*, (doctoral dissertation), Linköping: Linköping University.

Burns, T. and Stalker, G.M. (1961) *The Management of Innovation*, London: Tavistock Publications.

Bursche, J. (1986) 'State-of-the-art in the field of material requirements planning', in E. Szelke and J. Browne (eds) *Advances in Production Management Systems*, Amsterdam: North Holland Publications.

Caiden, N. and Wildavsky, A. (1974) *Planning and Budgeting in Poor Countries*, New York: John Wiley.

Campell, J. (1989) *Winston Churchill's Afternoon Nap*, London: Paladin Grafton Books.

CAPMAS (Central Agency for Public Mobilization and Statistics), Cairo.

Cooper, C. (1973) *Science, Technology and Development*, London: Macmillan.

Cooper, M. (1983) 'Egyptian state capitalism in crisis', in T. Asad and R. Owen (eds) *Sociology of Developing Countries: The Middle East*, London: Macmillan Press.

COPICS (Communication Oriented Production Information Control System). Volumes I–VII. IBM form numbers G320–1974 through G320–1980, New York: IBM Corporation.

COPICS för Material – och Produktionsstyrning (in Swedish), IBM form number GAM–0435, Stockholm: IBM Svenska AB.

COPICS Management Overview, IBM form number G320–1230, New York: IBM Corporation.

COPICS Advanced Function/Material Requirements Planning, IBM form number SB11–5531, New York: IBM Corporation.

Corke, D.K. (1977) *Production Control in Engineering*, London: Edward Arnold.

Daghistani, A. (1985) 'The race of development – Egypt vs. others', *L'Egypt Contemporaine*, Juillet, Le Caire.

Danielsson, A. (1983) *Företagsekonomi en Översikt* (in Swedish), Lund: Studentlitteratur.

Deyo, F. (1978) 'Local foremen in multinational enterprise: a comparative case study of supervisory role tensions in Western and Chinese factories in Singapore', *Journal of Management Studies*, 15.

Douglas, M. (1967) 'The Meaning of Myth', in E. Leach (ed.) *The Structural Study of Myth and Totemism*, London: Tavistock Publications.

Eilon, S. (1962) *Elements of Production Planning and Control*, New York: Macmillan.

El Dabaa, M. (1983) *Cost-benefit Analysis in the Developing Countries. A Case Study of a Caustic Soda Plant in Egypt* (dissertation paper), Copenhagen: Copenhagen School of Economics.

Eloranta, E. (1983) in E. Warman (ed.) *Computer Applications in Production and Engineering*, Amsterdam: North Holland Publications.

Ellul, J. (1964) *The Technological Society*, New York: Vintage Books.

Elzinga, A. and Jamison, A. (1981) *Cultural Components in the Scientific Attitudes to Nature – Eastern and Western Modes?*, Lund: Research Policy Institute.

Fleming, W. (1966) 'Authority, efficiency and role stress: problems in the development of East-African bureaucracies', *Administrative Science Quarterly*, 11.

Frank, A.G. (1978) *Dependent Accumulation and Underdevelopment*, London: Macmillan Press.

Galbraith, J. (1970) 'Environmental and technological determinants of organizational design', in J. Lorch and P. Lawrence (eds) *Studies in Organizational Design*, Homewood, Ill.: Irwin & Dorsey.

Girgis, M. (1977) *Industrialization and Trade Patterns in Egypt*, Kiel: University of Kiel.

Gupta, P.P. (1981) 'Policy framework for development of computer technology and applications', in J. Bennet and R. Kalman (eds) *Computers in Developing Nations*, Amsterdam: North Holland Publications.

Hall, E. (1959) *The Silent Language*, New York: Doubleday.

Handoussa, H. (1986) 'The South Korean success story: comparison and contrasts with Egypt', *L'Egypte Contemporaine*, Janvier, Le Caire.

Harris, R. and Kearny, R. (1963) 'The effects of political change on the role set of the senior bureaucrats in Ghana and Nigeria', *Administrative Science Quarterly* 13.

Hermele, K. (1982) *Swedish Auto Firms in Latin America* (mimeo), Lund: Research Policy Institute.

Hofstadter, D. (1980) *Gödel, Escher, Bach: An Eternal Golden Braid*, New York: Vintage Books.

Issawi, C. (1982) *An Economic History of the Middle East and North Africa*, London: Methuen.

Jamin, K. (1984) 'Die Bedeutung der Informations-und

Kommunikationstechnologie für Entwicklungsländer' (in German), Köln: Carl Duisberg Gesellschaft.

Janik, A. and Toulmin, S. (1973) *Wittgenstein's Vienna*, New York: Simon & Schuster.

Jansson, H. (1982) *Interfirm Linkages in a Developing Economy*, (doctoral dissertation), Uppsala: Uppsala University.

Kalman, R. (1981) 'Eight strategic issues for informatics', in J. Bennet and R. Kalman (eds) *Computers in Developing Nations*, Amsterdam: North Holland Publications.

—— (1982) *Comparative Study in Strategies and Policies for Informatics*, Paris: UNESCO.

—— (1983) 'Regional concepts – ten years after the UN report on applications of computer technology for development', in R. Kalman (ed.) *Regional Computer Cooperation in Developing Countries*, Amsterdam: North Holland Publications.

Katz, D. and Kahn, R. (1966) *The Social Psychology of Organizations*, New York: John Wiley & Sons.

Kaul, M. and Kwong, H.C. (1988) *Information Technology in Government – The African Experiences,* London: The Commonwealth Secretariat.

Kazarian, E. (1986) 'Den Islamiska Synen på Ekonomin' (in Swedish) *Ekonomisk Debatt*, 3.

Kiggunda, M., Jörgensen, J. and Hafsi, T. (1983) 'Administrative theory and practice in developing countries: a synthesis', *Administrative Science Quarterly*, March.

Kuhn, T. (1962) *The Structure of Scientific Revolutions*, Chicago, Ill.: The University of Chicago.

Lawrence, P. and Lorsch, J. (1967) *Organizations and Environment*, Boston: Harvard Business School.

—— and Spybey, P. (1986) *Management and Society in Sweden*, London: Routledge & Kegan Paul.

Lind, P. (1985a) 'The advent of computerization: South', in K. Elliot and P. Lawrence (eds) *Introducing Management*, London: Penguin Books.

—— (1985b) 'Some notes on production management systems in less developed countries', in E. Szelke and J. Browne (eds) *Advances in Production Management Systems*, Amsterdam: North Holland Publications.

—— (1986) 'Computers, myths and development', *Information Technology for Development*, 2.

—— (1988) *On the Applicability of Computerized Production in an Egypt Industry*, Stockholm: The Royal Institute of Technology.

Lorsch, J. and Lawrence, P. (eds) (1970) *Organizations Design*, Homewood, Illinois: Irwin & Dorsey.

Luckham, A. (1971) 'Institutional transfer and breakdown in a new nation: the Nigerian military', *Administrative Science Quarterly*, 16.

Malinowski, B. (1931) 'Culture', in *Encyclopaedia of the Social Sciences*, volume 4, New York.

March, J. and Simon, H.A. (1958) *Organizations*, New York: John Wiley & Sons.

Maruf, A. (1981) *Computerization in Malaysia – Problem and Issues*, mimeo, Kuala Lumpur: University of Technology Malaysia.

Moore, C.M. (1980) *Images of Development. Egyptian Engineers in Search of Industry*, Cambridge, Massachusetts: MIT Press.

Muller, M. (1979) 'Policies for planning and utilization of computers – full use vs. effective use', in J. Bennet and R. Kalman (eds) *Computers in Developing Nations*, Amsterdam: North Holland Publications.

—— and Rayfield, W. (1977) *Effective Computer Utilization in Developing Countries: Policy Questions and Implementation Plan*, Conference proceedings, August 1977, Bangkok.

Naranyan, S., Ramachandran, S. and Ramakrishnan, M. (1981) *Capacity Burden Forecasting*, Conference paper, New Delhi.

Narasimhan, R. (1984) *Guidlines for Software Production for Achieveing Software Self-Reliance in Developing Countries*, Wien: UNIDO.

Niehoff, A. (1959) 'Caste and industrial organization in North India', *Administrative Science Quarterly*, 3.

Olhager, J. and Rapp, B. (1986) *Production Environment and Manufacturing Planning and Control Systems* (paper), Departments of Business Administration (Stockholm) and Production Economy (Linköping).

Ouchi, W. and Maguire, M.A. (1975) 'Organizational control: two functions', *Administrative Science Quarterly*, December.

Pagels, H.C. (1988) *The Dreams of Reason*, New York: Simon & Schuster.

Pascoe, M. (1978) 'Dangers in computer policies', *Computer Monthly*, October.

Pfeffer, J. and Salancik, G. (1978) *The External Control of Organizations*, New York: Harper & Row.

Pizam, A. and Reichel, A. (1977) 'Cultural determinants of managerial behavior', *Management International Review*, 17.

Plossl, G. (1973) *Manufacturing Control*, Reston, Virginia: Reston Publications.

Pondy, L. and Mitrof, J. (1979) 'Beyond open systems models of organizations', in L. Cummings and B. Staw (eds) *Research in Organizational Behavior*, Greenwich, Connecticut: JAI Press.

Rada, J. (1980) 'Microelectronics, information technology and its effects on developing countries', in J. Berting, S. Mills and H. Wintersberger (eds) *The Socio-Economic Impact of Microelectronics*, Oxford: Pergamon Press.

—— (1983) 'A Third World perspective', in G. Friedrichs and A. Schaff (eds) *Microelectronics and Society*, New York: New American Library.

Rahim, S. and Pennings, A. (1987) *Computerization and Development in Southeast Asia*, Singapore: Kefford Press Ltd.

Siddiqi, M.N. (1976) 'Muslim economic thinking: a survey of contemporary literature', in K. Ahmad (ed.) *Studies in Islamic Economics*, Jeddah: International Centre for Research in Islamic Economics.

Siffin, W. (1976) 'Two decades of public administration in developing countries', *Public Administration Review*, 36.

Smircich, L. (1983) 'Concepts of culture and organizational analysis', *Administrative Science Quarterly*, September.

Smolik, D. (1983) *Material Requirements of Manufacturing*, New York: Van Nostrand Reinholt.

Starr, M. (1972) *Production Management Systems and Synthesis*, Englewood Cliffs, NJ: Prentice Hall.

Stevens, J. (1985) 'Purchasing', in K. Elliot and P. Lawrence (eds) *Introducing Management*, London: Penguin Books.

Stewart, F. (1978) *Technology and Underdevelopment*, London: Macmillan Press.

Streeten, P. (1981) *Development Perspectives*, New York: St.Martin's Press.

Sumanth, D. (1984) *Productivity Engineering and Management*, New York: McGraw-Hill.

Tantawy, D., Hamdi, A. and Kader, S. (1984) *Experience Gained Through the Implementation of Maintenance Management Project in the Egyptian Iron and Steel Industry*, mimeo, Cairo: Egyptian Iron and Steel Company.

Thompson, J. (1956) 'On building an administrative science', *Administrative Science Quarterly*, 1.

—— (1967) *Organizations in Action*, New York: McGraw-Hill.

—— (1974) 'Technology, polity and societal development', *Administrative Science Quarterly*, 19.

Thompson, V. (1964) 'Administrative objectives for development administration', *Administrative Science Quarterly*, 9.

Tibi, B. (1984) 'Management im Islam' (in German), *Entwicklung und Zusammenarbeit*, 3.

Timsit, G. (1976) 'Modèles administratifs et pays en voie de développement', *International Review of Administrative Science*, 42(4).

UNIDO (1983) *Industry in a Changing World*, ID/304, Vienna.

UNIDO (1983) *Problem of Software Development in Developing Countries*, IS.383, Vienna.

UNIDO (1984) *International Industrial Restructuring and the International Division of Labour in the Automotive Industry*, IS.472, Vienna.

UNIDO (1984) *ECWA Mission Findings: Arabization Efforts*, E/ECWA/ID/WG.8/3, Vienna.

UNIDO (1986) *Industrial Development Review Series: Egypt*, IS.637, Vienna.

UN National Accounts Statistics (1981) New York.

Uzari, M. (1976) 'Some conceptual and practical aspects of interest-free banking', in K. Ahmad (ed.) *Studies in Islamic Economics*, Jeddah: International Centre for Research in Islamic Economics.

von Wright, G.H. (1957) *Logik, Filosofl och Språk* (in Swedish), Stockholm: Bonniers.

—— (1986) Vetenskapen och Förnuftet (in Swedish), Stockholm: Bonniers.

Voris, W. (1966) *Production Control*, Homewood, Illinois: Richard Irwin.

Webster's New World Dictionary (1962), New York: The World Publishing Company.

Wittgenstein, L. (1922) *Tractatus Logico-Philosophicus*, London: Routledge.
Woodward, J. (1965) *Industrial Organization: Theory and Practice*, London: Oxford University Press.
—— and Reeves, T.K. (1970) 'The study of managerial control', in J. Woodward (ed.) *Industrial Organization: Behaviour and Control*, London: Oxford University Press.
World Automotive Market (1988) *Automobile International*, New York: Johnston International Publishing.
World Development Report (1982 and 1986), London: Oxford University Press.
Yalman, N. (1967) 'The raw: the cooked-nature: culture', in E. Leach (ed.) *The Structural Study of Myth and Totemism*, London: Tavistock Publications.
Youssef, S. (1979) 'Toward a strategy of management research in Egypt', in S. El Sayed (ed.) *Management Development in Egypt*, Cairo: The American University in Cairo Press.

Index